# Multiple Feature Object-Based Change Detection Technology for Very High-Resolution Remote Sensing Image

Qiang Chen · Yunhao Chen · Mingyi Du

# Multiple Feature Object-Based Change Detection Technology for Very High-Resolution Remote Sensing Image

测绘出版社
Surveying and Mapping Press

Springer

Qiang Chen
Department of Geomatics and Urban Spatial Information
Beijing University of Civil Engineering and Architecture
Beijing, China

Yunhao Chen
Department of Geography
Beijing Normal University
Beijing, China

Mingyi Du
Department of Geomatics and Urban Spatial Information
Beijing University of Civil Engineering and Architecture
Beijing, China

ISBN 978-981-95-3916-1 ISBN 978-981-95-3917-8 (eBook)
https://doi.org/10.1007/978-981-95-3917-8

Jointly published with Surveying and Mapping Press
The print edition is not for sale in China (Mainland). Customers from China (Mainland) please order the print book from: Surveying and Mapping Press.
ISBN of the Co-Publisher's edition: 978-7-5030-4387-1

Translation from the Chinese language edition: "Multiple Feature Object-Based Change Detection Technology for Very High Resolution Remote Sensing Image" by Qiang Chen et al., © Surveying and Mapping Press 2021. Published by Surveying and Mapping Press. All Rights Reserved.

The original submitted manuscript has been translated into English. The translation was done using artificial intelligence. A subsequent revision was performed by the author(s) to further refine the work and to ensure that the translation is appropriate concerning content and scientific correctness. It may, however, read stylistically different from a conventional translation.



This Springer imprint is published by the registered company Springer Nature Singapore Pte Ltd.
The registered company address is: 152 Beach Road, #21-01/04 Gateway East, Singapore 189721, Singapore

# Preface

High-resolution remote sensing images (spatial resolution $\leq$ 10 m) have become the most important basic data for conducting research in the field of geography and remote sensing. They can record richer surface features, making it possible to obtain more detailed surface coverage change information. The sharp increase in data information and image feature quantity poses new challenges to the change detection of high-resolution remote sensing images. Traditional pixel-based image analysis methods can lead to more "same object different spectrum, same spectrum different object" problems, while object-oriented image analysis methods (OBIA) can fully utilise the advantages of rich image object feature quantity/types. It has gradually become the mainstream approach for high-resolution remote sensing.

Due to the limitations of existing theories and methods, the following problems still exist in the research of object-based high-resolution remote sensing image analysis (OBCD): ① The research mainly focuses on a single spectral feature, without fully utilising the multi-feature advantages of high-resolution remote sensing images; ② The selection of segmentation parameters, the choice of feature combinations, and the acquisition of thresholds largely depend on empirical relationships, lacking a quantitative evaluation of image object feature differences; ③ The research often focuses on how to accurately extract changing objects, neglecting the distinction and recognition of multiple change types.

In response to the above scientific issues, this book uses object-oriented image analysis as a guiding principle, emphasises the advantages of multi-feature analysis in OBIA, and aims to meet the recognition of multiple change types. It conducts research on key image object technologies, aiming to propose a set of multi-feature change detection methods and detection patterns for high-resolution remote sensing image objects. For spectral feature changes and texture pattern changes, two typical multi-change type areas in Beijing are selected. Various high-resolution remote sensing data such as Worldview-2/3 (WV2/3), SPOT6, and Resource 3 (ZY3) are used to verify the methods proposed in this book. The main research conclusions of this book are as follows.

(1) Optimal selection of image segmentation parameters and consistency difference evaluation. Based on the generation of image objects through multi-scale segmentation, a multi-band segmentation quality index, *mean* NSQI, is constructed. The optimal segmentation parameters are obtained through layered screening, and the quality of the optimal segmentation parameters is evaluated by constructing geometric/quantity consistency difference evaluation indices. Experimental results show that this method of optimal selection and evaluation of segmentation parameters can effectively avoid subjective human interference and is applicable to various high-resolution remote sensing images. In addition, the research also shows that when the segmentation scale is less than 200, the consistency difference evaluation index maintains a reliable correlation with the segmentation scale, and the superposition of multi-temporal images will affect the optimal selection and evaluation of segmentation parameters.
(2) Improvement and verification of the image object multi-feature selection optimisation algorithm. The Genetic Particle Swarm Optimisation algorithm (GPSO) is applied to the multi-feature optimisation selection of image objects, and the Ratio of Mean to Variance (RMV) fitness function is chosen as the evaluation function to test the results of GPSO feature selection. The research shows that the feature optimisation algorithm based on GPSO can significantly improve the automation, intelligence, and computational efficiency of the multi-feature selection process. Compared with other feature selection algorithms, GPSO has certain advantages in convergence, global search capability, and parameter setting sensitivity. In addition, the sensitivity analysis results of the initial parameters of GPSO show that the number of features to be selected and the initial population size of the particle swarm will affect the accuracy, convergence, and efficiency of the algorithm's OBCD. A better initial parameter setting can significantly improve the efficiency and accuracy of the algorithm.
(3) Adaptive Change Vector Multi-Feature Change Detection Method and Pattern. In response to the issue of overly concentrated change vectors (CVA) and homogenised features in the Change Vector Analysis, an adaptive change vector analysis (SAW-CVA) multi-feature change detection method is proposed. This method calculates the intensity and direction of image object changes, introduces self-adaptive weights (SAW) in the standardisation step of each feature, and solves it in the form of polar coordinates. A "two-step" detection pattern for multi-feature changes in high-resolution remote sensing image objects is designed, which includes change object extraction based on threshold segmentation and change-type recognition based on unsupervised classification. The research results show that SAW-CVA has good applicability and accuracy in change detection and is superior to standard CVA in recognising fine change types. The overall accuracy of change detection in two types of typical land experiments reached 88.28% and 86.03%, respectively. In addition, the research also shows that similar multi-feature change detection algorithms can be expressed in this pattern, and the performance of each threshold segmentation algorithm in SAW-CVA is similar. The expectation-maximisation algorithm is more suitable for parameter estimation without prior knowledge.

This book, guided by OBIA, carries out research along the main line of "object generation—multi-feature selection—change detection" and systematically discusses the theoretical methods and technical difficulties of image segmentation, multi-feature selection and change detection in high-resolution remote sensing. It preliminarily solves the problems of multi-feature analysis utilisation and multi-change type recognition in high-resolution remote sensing change detection, providing new ideas for multi-feature change detection in high-resolution images.

This book has received the support of the National Natural Science Foundation and the Beijing Future City Design Innovation Research Centre project and is a summary of the results of related projects. During the writing of this book, Chen Qiang was responsible for writing the main chapters, Chen Yunhao, Wang Dandan, Xia Haiping, Li Kangning, and Li Yuanyuan participated in the related research work of this book, Cheng Qianhao was responsible for the collation and proofreading of the whole book, and finally Du Mingyi compiled the manuscript. Li Jing, Jiang Jinbao, Jiang Weiguo, Tang Hong, Zhou Lei and others have provided guidance and help for the relevant content of this book, and we would like to express our heartfelt thanks here.

Due to the author's level, there are inevitably inappropriate places in the book, and we sincerely hope that readers will not hesitate to criticise and correct.

Beijing, China

Qiang Chen
Yunhao Chen
Mingyi Du

# Synopsis

This book addresses scientific issues related to multi-feature analysis and multi-change-type recognition in high-resolution remote sensing image change detection. Guided by object-oriented image analysis, it integrates key technologies such as multi-scale image segmentation, consistency difference analysis, genetic particle swarm multi-feature optimisation selection, and adaptive weight change vector analysis. Following the research technical route of high-resolution remote sensing "object generation—multi-feature selection—change detection", it primarily explores an object-oriented high-resolution remote sensing image multi-feature change detection theory, method, and process.

This book can be used as a reference for university teachers, researchers, and postgraduates specialising in remote sensing science and geographic information systems.

## Abbreviations

| | | |
|---|---|---|
| CCA | Canonical Correlation Analysis | Canonical Correlation Analysis |
| CS | Cuckoo Search | Cuckoo Search |
| CVA | Change Vector Analysis | Change Vector Analysis |
| ED1 | Euclidean Distance 1 | Euclidean Distance based on Geometric Consistency Difference |
| ED2 | Euclidean Distance 2 | Euclidean Distance based on Quantity Consistency Difference |
| EM | Expectation Maximization | Expectation Maximization |
| FNR | False Negative Rate | False Negative Rate |
| FPR | False Positive Rate | False Positive Rate |
| FS | Feature Selection | Feature Selection |

(continued)

(continued)

| | | |
|---|---|---|
| GA | Genetic Algorithm | Genetic Algorithm |
| GCP | Ground Truth Point | Ground Truth Point |
| GEOBIA | Geography Object Based Image Analysis | Geography Object Based Image Analysis |
| GLCM | Grey Level Co-occurrence Matrix | Grey Level Co-occurrence Matrix |
| HCA | Histogram Curvature Analysis | Histogram Curvature Analysis |
| JMD | Jeffreys-Matusita Distance | Jeffreys-Matusita Distance |
| MAD | Multivariate Alteration Detection | Multivariate Alteration Detection |
| MNF | Minimum Noise Fraction | Minimum Noise Fraction |
| MRS | Multi-resolution Segmentation | Multi-resolution Segmentation |
| NNC | Nearest Neighbour Classifier | Nearest Neighbour Classifier |
| NSR | Number of Segments Ratio | Number of Segments Ratio |
| OA | Overall Accuracy | Overall Accuracy |
| OBCD | Object Based Change Detection | Object-Oriented Change Detection |
| OB-HMAD | Object based—Hybrid Multivariate Alteration Detection | Object-Oriented Hybrid Multivariate Change Detection |
| OBIA | Object Based Image Analysis | Object-Oriented Image Analysis |
| OR | Over-Segmentation Ratio | Over-Segmentation Ratio |
| PSE | Potential Segmentation Error | Potential Segmentation Error |
| PSO | Particle Swarm Optimization | Particle Swarm Optimisation Algorithm |
| QR | Quality Ratio | Segmentation Accuracy Ratio |
| RR | Robust Regression | Robust Regression |
| SFFS | Sequential Float Forward Selection | Sequential Floating Forward Selection |
| UR | Under-Segmentation Ratio | Under-Segmentation Ratio |
| VHR | Very High Resolution | Very High Resolution |
| WV-2/3 | Worldview-2/3 | — |
| NSQI* | Normalized Segmentation Quality Index | Normalised Segmentation Quality Index |
| meanNSQI* | — | Mean NSQI Across Multiple Bands |
| GPSO* | Genetic Particle Swarm Optimization | Genetic Particle Swarm Optimisation Algorithm |
| RMV* | Ratio of Mean to Variance | Ratio of Mean to Variance |
| $SA_c$* | Semi-Annulus of Changed Area | Semi-Annulus of Changed Area |
| SAW* | Self-Adaptive Weight | Self-Adaptive Weight |
| SAW-CVA* | Self-Adaptive Weight-Change Vector Analysis | Self-Adaptive Weight Change Vector Analysis |
| $SC_u$* | Semi-Circle of Unchanged Area | Semi-Circle of Unchanged Area |

Note: * indicates methods or abbreviations proposed or constructed in this book; — indicates undefined.

# Contents

# Chapter 1
# Introduction

This chapter elaborates on the research background and basis of this book, systematically reviews the research progress on high-resolution remote sensing multi-feature change detection at home and abroad, and the problems that need to be solved urgently, and thus brings out the purpose and content of the research. With the goal of object-oriented image analysis, and the main line of "object generation—multi-feature selection—change detection", the generation process of image objects and the optimal selection of segmentation parameters are studied, the problem of image object feature optimisation combination based on genetic particle swarm algorithm is analysed, the theoretical framework of adaptive change vector analysis is constructed, and the "two-step" mode of multi-feature change detection of image objects is proposed.

## 1.1 Research Background

### *1.1.1 High-Resolution Remote Sensing*

In recent years, with the rapid development of remote sensing sensors and earth observation technology, remote sensing images are gradually showing a "three high" trend of high spatial resolution, high spectral resolution, and high temporal resolution. Among them, with the continuous improvement of spatial resolution, sensors can obtain very detailed ground object information, and the ability to finely identify ground objects has reached a higher level, which also makes many subtle changes hidden between various ground units gradually discovered, thus greatly improving people's cognitive ability of ground detail changes. Generally, remote sensing images with a spatial resolution of less than 10 m are referred to as high-resolution remote sensing images (very high resolution, VHR).



Q. Chen et al., *Multiple Feature Object-Based Change Detection Technology for Very High-Resolution Remote Sensing Image*, https://doi.org/10.1007/978-981-95-3917-8_1

At present, various commercial imaging satellites have stepped into the "sub-meter" stage, Table 1.1 lists the main high-resolution remote sensing satellites on the domestic and international market.

Commercial high-resolution satellites have seen rapid development in the past decade, with panchromatic image resolution reaching up to 0.31 m (Worldview-3/4). Meanwhile, the spectral resolution of high-resolution images is gradually increasing, with Worldview-3's multispectral images reaching 28 bands. China's high-resolution remote sensing satellites have also made significant progress, launching a variety of high-resolution remote sensing satellites for different purposes in recent years, with the maximum spatial resolution reaching 1 m.

China's "National Medium and Long-term Science and Technology Development Plan (2006–2020)" includes the development of high-resolution remote sensing for Earth observation as one of the major special projects. In the coming years, China will launch a large number of high-resolution remote sensing satellites, covering everything from panchromatic and multispectral to hyperspectral, from optical imaging to radar imaging.

**Table 1.1** Main high-resolution remote sensing satellites at home and abroad

| Satellite | Spatial resolution | Country | Launch date |
|---|---|---|---|
| IKONOS | Panchromatic 1 m, Multispectral 4 m | USA | September 1999 |
| QuickBird | Panchromatic 0.61 m, Multispectral 2.44 m | USA | October 2001 |
| SPOT-5 | Panchromatic 2.5 m | France | May 2002 |
| GeoEye-1 | Panchromatic 0.41 m, Multispectral 1.65 m | USA | March 2006 |
| EROS-B | Panchromatic 0.7 m, Stereo Pair | Israel | April 2006 |
| Worldview-1 | Panchromatic 0.5 m, Multispectral 2 m | USA | September 2007 |
| Radarsat-2 | 3 m × 3 m C-band SAR | Canada | December 2007 |
| RapidEye | Multispectral 5 m | Germany | August 2008 |
| Worldview-2 | Panchromatic 0.5 m, Multispectral 1.8 m | USA | October 2009 |
| SPOT-6 | Panchromatic 1.5 m, Multispectral 6 m | Europe | September 2012 |
| Pleiades | Panchromatic 0.5 m, Multispectral 2 m | France | December 2012 |
| ALOS-2 | 1 m × 3 m L-band SAR | Japan | May 2014 |
| Worldview-3 | Panchromatic 0.31 m, Multispectral 1.24 m | USA | August 2014 |
| Satellite | Spatial Resolution | Country | Launch Date |
| Worldview-4 | Panchromatic 0.31 m, Multispectral 1.24 m | USA | November 2016 |
| Ziyuan III | Panchromatic (Fore/Aft View) 3.5 m, Multispectral 6 m | China | January 2012 |
| Gaofen-1 | Panchromatic 2 m, Multispectral 8 m | China | April 2013 |
| Gaofen-2 | Panchromatic 1 m, Multispectral 4 m | China | August 2014 |
| Gaofen-3 | 1 m × 1 m C-band SAR | China | August 2016 |
| Gaofen-6 | Panchromatic 2 m, Multispectral 8 m | China | June 2018 |

The characteristics of high-resolution remote sensing images mainly include (Gong et al. 2006): ① a significant increase in data volume due to the significant improvement in spatial resolution, resulting in more pixel numbers for remote sensing images covering the same area; ② a decrease in spectral resolution, although some high-resolution remote sensing images can now provide 8-band images, the number of usable spectral bands provided by the main high-resolution data is significantly less than traditional optical remote sensing images due to imaging technology; ③ richer feature information of the image's objects, the improvement in spatial resolution makes the geometric, spectral and texture feature information of the objects displayed in the image more obvious.

High-resolution remote sensing images can observe various detailed changes on the Earth's surface at a smaller spatial scale, and can be applied in various fields such as environmental monitoring (Li et al. 2016; Thapa et al. 2015), urban planning (Li 2014; Zhang et al. 2014), disaster reduction and prevention (Dong et al. 2016; Barnes et al. 2007; Wang et al. 2015), land registration change survey (Jia et al. 2012), precision agriculture (Calderón et al. 2014) and so on. Of course, the application of high-resolution remote sensing images in these fields cannot be separated from the object recognition, classification and change detection technology based on multiple image features, among which change detection plays an indispensable role in the geoscience application of high-resolution remote sensing and is one of the hot directions of high-resolution remote sensing image processing analysis (Hussain et al. 2013).

### *1.1.2 Object-Oriented Image Analysis*

In the application and practice of high-resolution remote sensing images, it has been found that the traditional pixel-based image processing methods have encountered many problems when applied to high-resolution images. For example, due to the increase in spatial resolution, the data volume has greatly increased, and the object information is more detailed, but the decrease in spectral resolution has led to more overlap in the changing spectral distribution in the image, resulting in a large occurrence of "same object different spectrum" and "same spectrum different object" phenomena, significantly reducing the statistical separability of the image spectral domain and bringing errors to object recognition, classification and change detection; secondly, a large amount of object detail information makes the pixel-based image processing results appear serious "salt and pepper" phenomenon, which is not conducive to the recognition and extraction of object information; in addition, how to balance texture structure and spectral information, how to effectively use the rich feature information of the image, these are tasks that pixel-based image processing methods cannot handle.

Since the 1990s, with the increase in high-resolution remote sensing image processing, the object-based image analysis method (OBIA) was born. Initially, it was only applied in parcel analysis and was called the per-field classification method

(Lobo et al. 1996). As computer image segmentation technology gradually became an important technical link in OBIA, it was later called the object-oriented image analysis method (Blaschke et al. 2000). With the continuous innovation of OBIA technology and methods, its conceptual theory has been continuously improved, and its application scope has gradually expanded to all areas of remote sensing image analysis.

In the theoretical development process of OBIA, Blaschke and Hay have repeatedly updated and improved the definition of OBIA. At the first OBIA conference in 2006, a group of scholars led by Blaschke summarized the theory of object-oriented, and for the first time proposed the definition of OBIA. They believe that OBIA is "a method dedicated to segmenting images into meaningful image objects and evaluating them through features such as spectrum, shape, and texture" (Hay et al. 2005).

Subsequently, at the second OBIA conference in 2008, Hay (2008), Blaschke and Lang (2008) suggested that OBIA should be considered a branch of Geographic Information Science, serving as a bridge between remote sensing and Geographic Information Science to some extent. To highlight its importance as a branch of Geographic Information Science, OBIA is defined as "a branch of geographic science dedicated to the development of automated image segmentation methods" (Hay et al. 2008).

In 2010, Hay and Blaschke further developed the definition of OBIA, introducing the concept of "Geo-intelligence" and explaining it as "Geospatial Content in Context" (Hay et al. 2010). This implies that OBIA is no longer confined to the environment of remote sensing image processing, but should also include concepts and principles of GIS and landscape ecology, with an increased emphasis on the role of multi-feature analysis in the OBIA process.

Currently, OBIA has become a new research paradigm in Geographic Information Science (Blaschke et al. 2014). The important challenge in the development of OBIA at this stage is how to transform OBIA into a way of thinking about geographic analysis rather than a technology. Although image segmentation is still an important part of OBIA, it is more often used as a means of extracting image object features, emphasising that OBIA is a branch of Geographic Information Science, not just a technology or method.

Throughout the development of OBIA theory, it can be seen that people's understanding of OBIA has gradually risen from a technical level to a scientific theoretical level. More people are gradually recognising OBIA as a way of thinking about geographic analysis and applying it in various research fields of remote sensing and GIS. Although OBIA is currently only a research paradigm and there is not yet a complete OBIA theoretical system and model, as a new branch of geographic science, OBIA is flourishing. In the future development process, how to deeply understand and implement the latest definition of OBIA is an important aspect of the application of OBIA in various fields of geographic science.

### 1.1.3 Object-Oriented Change Detection

The characteristics of fine object information in high-resolution remote sensing images determine its stronger ability to recognise object changes than other remote sensing images, but it also brings many challenges to accurate change detection. Among them, the problems caused by image feature information related to the use of multi-feature analysis, and the recognition of multiple types of changes in real change detection scenarios, are the main issues in current high-resolution remote sensing change detection research. Change detection based on high-resolution remote sensing images can more effectively identify the feature changes and detailed information of different land cover types. However, in high-resolution remote sensing change detection, the redundancy of object features and the complexity of object details make it difficult to identify, becoming a problem in high-resolution remote sensing image detection, leading to poor detection accuracy and low efficiency in traditional change detection theories and techniques. High-resolution remote sensing images have multiple types of features such as spectrum, texture, geometry, and edge (Hao 2015), and the types of changes driven by different features are significantly different. The detection of changes in multiple types such as spectrum, texture, and geometry has gradually become the main research direction in current change detection research.

In current high-resolution remote sensing change detection research, the main method used is object-oriented change detection (Blaschke et al. 2014), which focuses on extracting change areas. However, some scholars (Du Peijun et al. 2012; Lv 2014; Zhang et al. 2017) believe that high-resolution remote sensing change detection should fully utilise the rich multi-feature information of high-resolution remote sensing images. While quickly and accurately extracting change areas, it should be guided by the multi-features of image objects to identify or distinguish different types of changes in these change objects, to meet the application needs in various real scenarios. Object-based change detection (OBCD) research is an important research category in OBIA (Blaschke et al. 2014), and multi-feature analysis has become the main means to solve the problem of high-resolution remote sensing image change detection.

On the one hand, traditional pixel-based change detection methods cannot effectively utilise the multi-feature advantages of high-resolution remote sensing images, and the complex and detailed object information instead brings more errors to the results, making the algorithm inefficient and unreliable (Hussain et al. 2013). OBCD can take advantage of multi-feature analysis, but current research still lacks distinction and selection of its various features. The impact of different combinations of multi-features for different research objects also needs attention.

On the other hand, change detection is no longer just about extracting change objects, and there is an increasing demand for the identification and distinction of multiple types of changes (Tewkesbury et al. 2015). The requirements for change detection vary in different scenarios. For instance, in compact urban spaces, due to the complex spatial relationships of geographical features, there are many pseudo-changes caused by building shadows and sensor angle deviations. The single spectral property change can no longer fully reflect the situation, and it is necessary to use texture patterns and geometric structures to identify these pseudo-change objects. Similarly, for a piece of agricultural land, different farming methods (mainly different planting directions) cannot be reflected in spectral change types, and other types of changes need to be identified. Generally, the main types of changes include spectral changes, texture pattern changes, and geometric structure changes. Considering that high-resolution remote sensing images have various features such as spectrum, texture, geometry, and spatial relationships, and for each image object, the change types dominated by different features vary. These differences can fully explain the main reasons for the changes, and multi-type change detection is based on these feature differences, revealing the dominant change features, thereby identifying different types of changes. Therefore, how to accurately and efficiently identify the types of changes driven by different dominant features is a hot and difficult point in current high-resolution remote sensing change detection research.

At present, the problems of traditional high-resolution remote sensing change detection not being able to accurately and efficiently identify various types of changes are specifically reflected in: (1) Not fully utilising multi-features of images: traditional change detection methods cannot utilise the rich features of image objects, and such change information is often expressed through multi-dimensional feature space. Single features have limited ability to identify their changes, cannot accurately express the change information of the object, and the results of its change detection are biased (Leichtle et al. 2017); (2) There are many redundant features that interfere: high-resolution image features are large and complex, there are many useless redundant feature information, so that it is difficult to determine the dominant change features of each image object during the change detection process, resulting in low algorithm efficiency (Hussain et al. 2013); (3) Low separability of change features: conventional change detection methods, when using multi-features, the feature standardisation methods used will make the change feature vectors too concentrated, with low separability, which is not conducive to the identification and classification of multiple types of changes, so it needs to be adaptively improved.

To address the above issues, the following solutions are considered. Firstly, the rich pixel information of high-resolution remote sensing images brings many disadvantages to pixel-based image analysis methods, and object-oriented methods can overcome these disadvantages. However, for the generation of image objects, it is necessary to strictly ensure that image objects at different times are comparable, and that image objects have high internal homogeneity and external heterogeneity, which places high demands on the selection of image segmentation methods and parameters; secondly, the multi-feature information of image objects is large, and there is also a lot of redundant information, and there is mutual interference, so it is necessary to select the appropriate feature combination. In the selection process, try to use intelligent and highly automated preferred methods to reduce human subjective interference and ensure the reliability and efficiency of the algorithm; finally, it is necessary to improve the traditional OBCD method, through adaptive feature weighting, to improve the separability of change objects, and perform vector calculation in the polar coordinate system to achieve the identification of multiple types of changes.

Therefore, this book focuses on the scientific issues in the multi-feature analysis of high-resolution remote sensing change detection, and combines the typical land use types in the Beijing area, guided by object-oriented image analysis, to explore and verify the feasibility of the "object generation—multi-feature selection—change detection" technical route. Specifically, image segmentation is used as the basis to generate image objects as the basic processing and analysis units, analyse the characteristics of image object features, select the most suitable feature combination as the input for change detection, and focus on the study of change detection methods based on multi-feature analysis of image objects, and explore methods that can achieve multiple types of change recognition.

## 1.2 Research Progress at Home and Abroad

Regarding the problems of underutilisation of multiple features, interference of redundant features, and low separability of features in high-resolution remote sensing change detection, scholars at home and abroad have relevant research foundations and have proposed many solutions, which can be summarised as optimisation of image segmentation, optimisation of feature selection, and optimisation of remote sensing change detection methods. The following will be summarised, summarised, and analysed separately.

### *1.2.1 Image Segmentation*

The generation of VHR remote sensing image objects mainly relies on image segmentation technology, which is the basis of object-oriented image analysis methods (Tong et al. 2012; Duro et al. 2012). Image segmentation is a technique that divides images into objects with different characteristics to extract target objects (Li et al. 2013; Lang et al. 2010).

An important premise of object-oriented segmentation is to construct image objects with consistent features (including spectral, texture, shape, etc.), and to use objects as the basic operation units for image analysis. In OBIA, image segmentation is usually the most critical step. Only by obtaining better segmented objects can better results be achieved in object-oriented information extraction, target recognition, and change detection of multi-temporal remote sensing images (Blaschke et al. 2014). Image segmentation algorithms can generally be divided into three types: edge detection-based segmentation algorithms (Tian et al. 2007), region-based segmentation algorithms (Evans et al. 2002), and threshold-based segmentation algorithms. Among the region segmentation algorithms, the Multi-Resolution

Segmentation (MRS) method (Wang 2007; Baatz et al. 2000) has been proven to be a universally applicable segmentation algorithm in the field of remote sensing applications (Neubert et al. 2007). It can generate image objects with more geographical significance and has features such as strong adaptability and user customisation (Hay et al. 2003; Marpu et al. 2010).

The main parameters of MRS include segmentation scale, shape heterogeneity weight, and smoothness weight. The choice of segmentation scale directly relates to the results of subsequent image analysis. If the segmentation scale is too large, resulting in too small segmentation objects, smaller objects will be "submerged" by larger objects, which is an under-segmentation situation. If the segmentation scale is too small, it will cause the segmentation objects to be too "fragmented", which is an over-segmentation situation. In these two situations, the boundaries and numbers of image objects cannot be consistent with the actual geographical objects, affecting the accuracy of change detection (Yu et al. 2010), as shown in Fig. 1.1.

The current optimal segmentation scale selection is divided into supervised discrimination and unsupervised discrimination (Yu et al. 2010; Zhang et al. 2008). Supervised discrimination involves conducting "trial-and-error" experiments at different segmentation scales to obtain different segmentation results, and then relying on manually stored knowledge for visual discrimination (Myint et al. 2011), comparing the inconsistency or dissimilarity between the segmentation results and the reference results. This method does not require much theoretical support, is practical, but its strong subjectivity makes it difficult to obtain objective segmentation results. Many scholars (Carleer 2010; Costa et al. 2008; Möller et al. 2007; Tian et al. 2007; Trias-Sanz 2005) use supervised discrimination to select the scale,

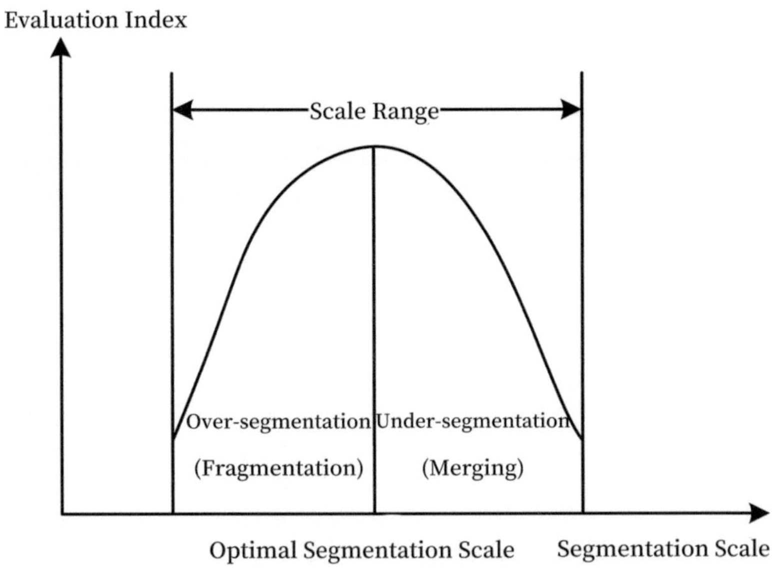


**Fig. 1.1** Relationship between segmentation scale and evaluation index

among which Tong proposed a parameter optimisation model based on fuzzy theory, measuring the segmentation quality by the fuzzy membership degree of the segmentation object partially belonging to the target object (Tong et al. 2012); Liu constructed a segmentation quality evaluation standard (Euclidean Distance 1/2) by analysing the dissimilarity and inconsistency between the segmentation result object and the reference object (Liu et al. 2012); Witharana, based on Liu's work, used the Krusal-Wallis test and Wilcoxon signed-rank test and other non-parametric tests to analyse the sensitivity of the empirical model, thereby verifying the reliability of the evaluation index (Witharana et al. 2014).

Unsupervised discrimination generally refers to the construction of some robust, highly adaptive indices or indicators to measure segmentation quality based on the characteristics of the image object itself, with less human interference and user participation (Johnson et al. 2011; Martha et al. 2011). Some scholars (Drăguţ et al. 2010; Espindola et al. 2006) have proposed some quantitative discrimination methods, such as: constructing an area matching index to measure the matching degree of segmentation objects (Neubert et al. 2007), or evaluating the quality of segmentation results through boundary coincidence (Prieto et al. 2003). Compared with supervised discrimination methods, unsupervised discrimination methods have stronger theoretical and complexity, but are not conducive to popularisation and promotion. Therefore, this book uses the multi-scale segmentation algorithm as the basic means of generating image objects, combines supervised and unsupervised discrimination methods, uses the index construction method in unsupervised methods as the method of segmentation parameter optimisation, and uses the consistency difference analysis method in supervised methods as the method of segmentation object quality evaluation.

### *1.2.2 Multi-feature Selection*

In the process of object-oriented change detection, by selecting the features of the object, redundant feature information can be eliminated, data dimensionality can be reduced, and the complexity of the algorithm can be lowered. Reasonable feature selection can maximise the retention of change information of image object features, improving the efficiency of change detection.

Feature selection is based on specific evaluation criteria, selecting the subset of features from the original feature space that best meets these criteria. Since the 1960s, research on feature selection problems has been carried out in the fields of statistics and machine learning, and has been continuously improved and innovated (Chen et al. 1997; Kira et al. 1992; Kononenko 1994), until 1997 when Dash and Liu proposed the basic algorithm framework for feature selection (Dash et al. 1997), which mainly includes subset generation, subset evaluation, stopping criteria, and result validation (Yu et al. 2003). Current research on feature selection problems mainly focuses on the stages of subset generation and subset evaluation. Subset generation mainly relies on the choice of search strategy, while subset evaluation relies on the choice of evaluation criterion function. Therefore, feature selection methods are developed through search strategies and evaluation criteria.

1. Feature selection methods divided based on search strategy

Feature selection methods divided based on search strategy generally have three types: complete search, heuristic search, and random search (Sun et al. 2004).

Complete search mainly consists of exhaustive search and branch and bound methods. The exhaustive search method involves a one-by-one search of the entire feature space, which is computationally intensive and inefficient. The branch and bound method (Narendra et al. 1977) represents the search space of the algorithm through a tree, with the root of the tree representing the feature set, and searches from the root for the required feature subset (Chen 2003). Compared with the exhaustive search, the branch and bound method reduces time cost, but it is difficult to define the number of features in the candidate feature subset and design a monotonic evaluation function. Heuristic search mainly includes individual optimal feature combination, sequential forward selection, sequential backward selection methods, plus l minus r selection methods, floating search algorithms, etc. The individual optimal feature combination ranks features based on the evaluation function values of each individual feature, selecting the first few features that meet the conditions as the optimal features (Rakotomamonjy 2007). The Sequential Forward Selection (SFS) method (Marcano-Cedeño et al. 2010) is a "bottom-up" search method, which adds a feature that optimises the evaluation function value each time, starting from an empty feature subset. The Sequential Backward Selection (SBS) method (Peng et al. 2014) is a "top-down" search method, starting from the full feature set, and eliminating one feature at a time to optimise the evaluation function value after elimination. The plus l minus r selection method combines the SFS and SBS methods. When $l > r$, it is a "bottom-up" algorithm, using SFS to add l features to the current feature set, and then using SBS to eliminate the r worst features. When $l < r$, it is a "top-down" algorithm, starting from the original feature set, using SBS to eliminate r features, and then using SFS to add l features until the requirements are met (Feng et al. 2011). The floating search algorithm changes the fixed l and r approach, adopting a floating step length to obtain the optimal feature subset (Tan et al. 2014; Ververidis et al. 2008). Although these methods have high search efficiency, they abandon the ability to search globally and are prone to falling into local optimum search traps. This book uses the SFS algorithm as a comparison algorithm for multi-feature selection, to analyse its differences with the random search algorithm. Random search refers to each feature subset to be searched being randomly generated, with a defined threshold, and multiple iterations of search, to obtain the optimal feature subset in the entire search process. Feature selection is essentially a combinatorial optimisation problem, so some evolutionary computation algorithms can be used to solve this problem. Evolutionary search maintains good randomness, strong global search ability, and common evolutionary algorithms include simulated annealing algorithm (Brusco 2014), genetic algorithm (Anusha et al. 2015; Zhu et al. 2007), differential evolution algorithm (Neri et al. 2010), backtracking search optimisation algorithm (Civicioglu 2013), cuckoo search algorithm (Civicioglu et al. 2011; Yang et al. 2010), ant colony optimisation algorithm (Chen et al. 2010) and particle swarm optimisation algorithm (Eberhart et al. 2001). Usually, these algorithms use the classification performance of the feature subset on the training set as the evaluation criterion for the feature subset, so these algorithms can solve the problem of local convergence well. However, due to the dependence of the algorithm on random factors, the results are highly unstable and experimental results are difficult to reproduce. If the number of feature inputs is too large, the efficiency of the algorithm decreases. This book focuses on the application of the genetic particle swarm optimisation algorithm (GPSO) in the process of image object feature selection, and discusses the reliability, convergence and algorithm efficiency of the GPSO feature selection algorithm in multi-feature change detection applications.

2. Feature selection methods based on evaluation criteria

Feature selection methods based on evaluation criteria can basically be divided into two categories: filter and wrapper (Chuang et al. 2011), and different feature selection methods will determine different discriminant functions (Fig. 1.2). The filter method selects features based on the nature of the data, based on a specific evaluation criterion, and the feature selection process is separate from the change detection process. The wrapper method applies a learning algorithm to each input feature subset, and then evaluates the input feature subset through the evaluation function of the fused change detection result, and the feature selection process is parallel to the change detection process. The filter method uses an evaluation function or criterion to evaluate the predictive ability of features. The general evaluation functions can be divided into distance measurement, information measurement and dependency measurement (Yao et al. 2012). Distance measurement is generally a measure of the separability, difference and recognisability between two or more categories. Common distance measures include Euclidean distance, Mahalanobis distance (Bruzzone et al. 2010), Bhattacharyya distance (Xuan et al. 1996) and between-class and within-class distance (Sun et al. 2004). Information measurement usually uses information gain or mutual information for measurement. Common information measurement methods include BIF (Best Individual Feature (Jain et al. 2000), mRMR (minimal-Redundancy and Maximal-relevance) (Peng et al. 2005), and FCBF (Fast Correlation-based Filter) (Yu et al. 2004) are among the methods used. In recent years, a large number of feature selection methods based on information entropy have emerged (Qu 2010; Zhao et al. 2009; Berrendero et al. 2015; Gao et al. 2013). Although the evaluation criteria vary, the core idea is to find the feature subset with the highest relevance to the category. Dependency measurement is characterised by some statistical correlation coefficients to represent the degree of separability of features relative to the category. For example, Ding et al. (2011) used the F-test and Pearson correlation coefficient in mRMR to measure the degree of correlation between features and between features and categories, thereby achieving continuous feature selection. Wei used the square correlation number as the discriminant function of his forward orthogonal search algorithm to measure the independence between features and achieve data dimension reduction (Wei et al. 2007). Mitra used the least square regression error as a measure of feature similarity index, as the discriminant function of unsupervised feature selection algorithm, to achieve data dimension reduction (Mitra et al. 2002). In addition, commonly used metrics include probability error, t-test (Nina et al. 2008) and Fisher criterion (Wang et al. 2011). Filters can select a superior feature subset based on appropriate feature evaluation criteria/functions. The obvious advantage is that it can quickly eliminate noise and redundant features, reduce the search space, and is efficient and versatile. However, it cannot guarantee that the size of the selected feature subset is minimal, and there will be some interfering features. In addition, filter feature selection methods also include some feature dimension reduction methods such as principal component analysis (Morchid et al. 2014), independent component analysis, etc., to remove some interfering features and highlight the main features.

The wrapper method incorporates the feature algorithm as part of the machine learning algorithm, using classification performance as the evaluation standard for feature importance. It trains the learner with the selected feature subset and evaluates the selected feature subset based on the performance of the test set on the learner. Although it is less efficient than the filter method, the size of the feature subset it selects is relatively small, which is conducive to the identification of key features. Ashoori et al. (2011) proposed using the method of classification regression tree to screen the input items of the neural network model, determine the optimal feature set based on the prediction error rate, and input it into the neural network model for prediction. Yao Dengju used the classification accuracy of the random forest classifier as the criterion for feature separability, and ranked the importance of features based on the variable importance of the random forest algorithm itself (Yao et al. 2014).

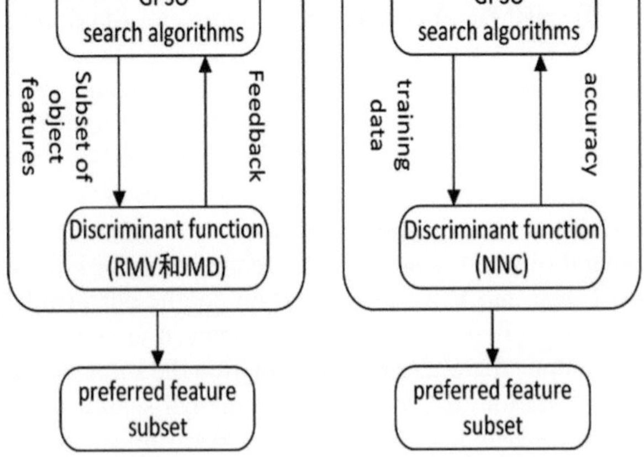


**Fig. 1.2** Schematic diagram of filter and wrapper feature selection methods

Chapran used the classification error rate of the minimum distance classifier as the standard for evaluating input features, thereby achieving feature selection (Chapran 2012). Shima used the SVM classifier as the machine learner of the wrapper, and selected the optimal feature subset by evaluating the classification error rate (Shima et al. 2004).

This book will use dependency measurement as the main method of the discriminant function, focusing on how to achieve the distinguishing ability of features for different categories, and build a fitness function based on the ratio of mean-value to variance (RMV). In addition, it also uses the Jeffreys-Matusita distance (JMD) filter method and the nearest neighbour classifier (NNC) wrapper method to compare with the RMV fitness function.

### *1.2.3 Change Detection*

In 1961, Rosefield was the first to use periodic remote sensing images to detect changes in surface cover. Change detection generally refers to a technique of quantitatively analysing data from different periods to extract surface change information (Chen et al. 1990).

Over the past few decades, a large amount of research has been done on the technical methods and applications of land use/cover change detection from different perspectives. The basic process of remote sensing change detection has been summarised: image preprocessing, change information discovery, change area extraction and change type distinction (Zhang et al. 2005), where the key steps are "change area extraction" and "change type distinction". Current research on remote sensing change detection is mostly from the comparison of two-phase images, and a small amount of research is on the comparison of long time series images. This book mainly discusses the methods and applications of two-phase change detection.

Based on previous research (Coppin et al. 2004; Hussain et al. 2013; Lu et al. 2004), object-oriented change detection methods are divided into post-classification comparison, image algebra, direct classification, image transformation, and change vector analysis.

(1) The Post-Classification Change method first classifies images from different periods separately, obtains the transition matrix of each category, and then obtains the change information of each category. This method has been used for Landsat classification images (Berrendero et al. 2015; Blaschke 2010) and high-resolution remote sensing images (Barnes et al. 2007; Blaschke 2005a). Coulter and Tescher used the post-classification comparison method to identify changes in urban buildings (Coulter et al. 2011; Tescher 1988). This method requires comparing the differences between two complete classification images, which are often obtained through different training data and methods. The differences in classification can introduce more inconsistencies to change detection. The main drawback of the post-classification comparison method is that it heavily

relies on the quality of the classification images, which can easily lead to error accumulation.

(2) Image Arithmetic includes image differencing and image ratioing. Subtracting or dividing corresponding bands of images from different times is known as image differencing or ratioing. This method can intuitively present the radiance differences described by the images, which can reflect changes in the land cover. The advantage of this method is that it is simple and quick to implement. However, it is difficult to determine the change threshold, and the difference (or ratio) images contain a lot of noise and cannot provide the type of change. To add more thematic information to the difference images, the image radiance is usually obtained through band operations to get vegetation indices or other thematic coverage image information, and then the change information is obtained through image arithmetic. For example, Coulter used different regional measurements of NDVI to detect changes in vegetation cover (Coulter et al. 2011); Tewkesbury used image ratioing on aerial remote sensing images to identify vegetation degradation (Tewkesbury et al. 2015). Cases of extracting urban impervious surface cover changes through image band operations to monitor urban dynamic changes are more common (Dams et al. 2013; Gangkofner et al. 2010; Lu et al. 2011), including those that measure changes through difference images obtained from image spectral indices and band operations (Jin et al. 2013). In summary, image arithmetic is suitable for rapid change detection studies, but it cannot compare multiple features (bands), and the changes in the images do not fully reflect the actual spectral changes.

(3) Direct Classification involves stacking multi-temporal images in bands, then directly performing supervised or unsupervised classification on them to obtain land cover category information and category transition information. The advantage of this method is that it can obtain category change information through the classification process, so many scholars use this method to study forest changes. Hame used an unsupervised classification method to obtain tree change information from multi-temporal images (Hame et al. 1998). For relatively complex land cover type changes, if multiple types of changes are of interest, supervised machine learning methods or decision tree methods are generally used. For example, Schneider used SVM and decision tree classification methods to obtain changes in various land cover types in urban areas from multi-temporal Landsat images (Schneider 2012); Gao used the Classification and Regression Tree (CART) method to extract changes in all types of land cover (Gao et al. 2012). Direct classification generally requires training samples. If the number of samples is too small, the efficiency of machine learning or decision classification models is poor, which greatly affects supervised classification methods; for unsupervised classification methods, only obvious changes in a single type can be obtained, and some fine changes cannot be detected or identified (Blaschke et al. 2014).

(4) The most commonly used method in Image Transformation is Principal Component Analysis (PCA) (Byrne et al. 1980) and Multivariate Alteration Detection (MAD) (Bruzzone et al. 2000b). Both compress the related information between

bands, highlight and extract the main difference information, and achieve change detection. Deng used PCA on multi-temporal Landsat and SPOT5 images, then performed supervised classification on the PCA images to obtain change information (Deng et al. 2008); Doxani found that using MAD transformation on VHR remote sensing images makes the change information more prominent (Doxani et al. 2012); Listner also used MAD transformation in an object-oriented processing manner to achieve change detection (Listner et al. 2011). However, they all found that if there is a high correlation between the input features, MAD transformation will have mathematical instability, so PCA transformation can be used before MAD transformation, and the first few principal components can be used as MAD input features (Angeline 1998; Anusha et al. 2015); MNF transformation can also be used on the results of MAD transformation, so that the change information is more concentrated, increasing the separability of the features (Ashoori et al. 2011; Baatz et al. 2000). However, in practical applications, the number of PCA input principal components is a problem. Too few will miss change features, and too many will cause data redundancy. Transformation methods can input multiple features for change detection, but how to determine the number of input features is a difficult problem, and transformation methods can only identify changes, not distinguish types of changes.

(5) Change Vector Analysis (CVA) was originally designed by Malila for changes in two-dimensional spectral space at two different times. It uses change vectors (CV) to mine the change information of two remote sensing images at different times (Malila 1980). This change vector is described by two elements in two-dimensional space: change intensity and change direction. These two elements can identify the type of change that occurred between the two time points. Specifically, the change intensity reflects the degree of change in image radiance or other thematic features, while the change direction reflects the type of change (Tewkesbury et al. 2015; Zhuang et al. 2016) Change Vector Analysis (CVA) can avoid the cumulative errors brought about by image classification in each period, which are common in other change detection methods such as post-classification comparison. Moreover, it can utilise more bands for analysis. When dealing with multispectral image change detection, CVA transforms it into a feature vector of dual spectral channels for analysis. By subtracting the feature vectors corresponding to the two images, a change vector rich in change information can be obtained. To date, CVA has been widely used in many fields of remote sensing change detection (Allen et al. 2000; Bovolo et al. 2007; Bovolo et al. 2008; Bovolo et al. 2012; Chen et al. 2003; He et al. 2011; Johnson et al. 1998; Maeda et al. 2011; Nackaerts et al. 2005; Thonfeld et al. 2016; Warner 2005; Xiaolu et al. 2011; Yunhao et al. 2002).

(6) Object-Based Change Detection (OBCD). In recent years, object-oriented change detection research has made significant progress. Bruzzone and Prieto proposed a parcel-based unsupervised change detection method (Bruzzone et al. 2000a); Willhauck et al. conducted object-oriented change detection research on

land use changes using aerial photos and SPOT remote sensing images, demonstrating a high degree of automation in the entire change detection process (Willhauck et al. 2000); Civco compared other change detection techniques with object-oriented change detection techniques, including post-classification methods, neural networks, cross-correlation analysis, and knowledge decision based on expert systems, concluding that object-oriented techniques have great potential (Civco et al. 2002); Walter used a GIS database as a basis, combining pixels into existing objects in the GIS database, each object being an n-dimensional feature vector, and classified them in a supervised manner (Walter 2004); He used object similarity to obtain land use/cover change information (He et al. 2005); Blaschke proposed a GIS framework theory, classifying shape differences of the same object in different datasets, and applied this framework to explore changes in object shape and thematic features in multi-temporal land use changes (Blaschke 2005b); Dronova used object-oriented and supervised classification methods to conduct wetland change detection research in the Poyang Lake area, proposing a semi-automatic training sample selection model and evaluating accuracy through fuzzy classification (Dronova et al. 2011).

### *1.2.4 Issues to Be Addressed*

In summary, high-resolution remote sensing change detection has always been a popular research field in geographical science. The identification of multiple types of changes is a scientific issue of great interest to scholars, thus promoting the development of techniques and methods such as image segmentation, feature selection, and remote sensing change detection. Although there have been many research achievements in image segmentation, feature selection, and change detection methods, which provide references for our next step of research, the following issues still exist:

(1) Current OBCD research mostly focuses on single spectral features, not fully utilising the advantage of multiple features in OBIA. Image objects have rich features, and their change information is often expressed through multi-dimensional feature space. Single features have limited ability to discriminate their changes, cannot fully express the change information of objects, and the results of change detection are biased. Traditional pixel-based change detection methods cannot utilise multiple features for analysis, but OBCD methods can take advantage of multiple features. However, most current research on OBCD still only focuses on the changes in the spectral features of image objects, without considering the recognition ability of other features for change information, which does not meet the requirements of OBIA for multi-feature analysis.

(2) The degree of automation and intelligence in the selection of segmentation parameters, the choice of feature combinations, and the acquisition of thresholds is low, affecting the efficiency and objectivity of OBCD. In the three stages of image segmentation, multi-feature selection, and change detection, parameter optimisation, weight setting, and threshold acquisition are involved. Many

documents have used various intelligent algorithms and evolutionary algorithms to optimise these steps, but they all require human intervention to some extent, which cannot balance the efficiency and objectivity of the algorithm. In addition, there is less research on the differences in different features of image objects in these processes. For example, in multi-feature change detection, the same threshold is often given to each feature or each object, ignoring the differences in features, and cannot reflect the advancement and comprehensiveness of multi-features compared to single features.

(3) Research mainly focuses on how to accurately extract change objects, neglecting the distinction and identification of multiple change types in image change objects. Traditional OBCD research mostly discusses how to quickly and accurately identify and extract change areas or objects. However, with the extensive use of high-resolution remote sensing images, the demand for distinguishing and identifying different types of changes in change objects is increasing, especially for some special areas, simply extracting change areas cannot fully explain the reasons for these changes. Therefore, on the basis of accurately extracting change objects, further research is needed on how to distinguish and identify different types of changes, so as to reveal the main reasons for the changes.

(4) There is a lack of a universally applicable theoretical framework and detection model for multi-feature OBCD research. Current research on each link of object change detection in high-resolution remote sensing images often operates independently, with a plethora of methods emerging, each with different applicable conditions and data requirements, and no unified theoretical framework has been formed. Therefore, it is necessary to summarise a universally applicable theoretical framework and detection model for multi-feature change detection of high-resolution remote sensing image objects based on these technologies, so that various similar methods can be uniformly expressed under such a theoretical framework and solve various common problems in image segmentation, feature selection, and change detection under its guidance.

## 1.3 Research Content

### *1.3.1 Research Content*

This book focuses on issues related to multi-feature analysis and change type identification in high-resolution remote sensing image change detection, based on the concept of object-oriented image analysis. The main research content includes the following.

(1) Generation of image objects based on multi-scale segmentation and optimal selection of segmentation scale. ① Discuss the image segmentation process based on the multi-scale segmentation algorithm; ② For the optimisation selection problem of various parameters in the algorithm, propose a segmentation

quality index based on the homogeneity within the image object and the heterogeneity outside, and automatically select the optimal segmentation parameters in a hierarchical screening manner; ③ Analyse the geometric and quantity consistency differences between the image segmentation object and the reference object, propose an image segmentation quality evaluation standard based on the consistency difference index, and evaluate the applicability of the optimal parameters.

(2) Multi-feature selection of image objects based on the genetic particle swarm optimisation algorithm. ① On the basis of image objects, extract the main features of the image objects, and for the multi-feature optimisation problem of image objects, propose a method for optimising the selection of multiple features of image objects based on the genetic particle swarm optimisation algorithm; ② On the basis of feature selection theory and the genetic particle swarm optimisation algorithm, combined with the fitness function based on the mean and variance ratio, explore the multi-feature optimisation method based on the genetic particle swarm; ③ Apply the algorithm to the research of multi-feature image object change detection, combine with other feature selection methods, and analyse the convergence, effectiveness, and efficiency of the algorithm in the application of multi-feature image object change detection.

(3) Multi-feature change detection of image objects based on adaptive weight change vector analysis. ① For the problem that traditional change detection methods cannot or are not easy to identify change types, improve the change vector analysis method, combine the calculation of change intensity and change direction, and construct a multi-feature change detection theoretical framework based on adaptive weight change vector analysis; ② Under the support of this framework, a "two-step" detection model for multi-feature changes of high-resolution image objects was designed; the expectation–maximisation algorithm is used to extract change objects, and the $K$ -means algorithm is used to identify different change types; ③ Discuss the compatibility of this detection model with other traditional change detection methods, and analyse the impact of different threshold segmentation algorithms on the process of extracting change objects.

### *1.3.2 Technical Route and Chapter Organisation*

This book follows the research technical route of "object generation - multi-feature selection - change detection" (Fig. 1.3), selecting two research areas of typical land cover change in the Beijing area. Firstly, the multi-scale segmentation algorithm is used for image segmentation, and the meanNSQI index is constructed and the optimal segmentation parameters are selected by hierarchical screening; the consistency difference between the reference object and the segmented object is compared to evaluate the segmentation parameters. Secondly, the main features of the image objects are extracted to build an image object feature library; the genetic particle swarm optimisation algorithm is used, combined with the RMV fitness function, to optimise the multi-feature selection of image objects, and the convergence, reliability, and efficiency of the algorithm are compared and analysed. Finally, according to the different dominant features of each object, adaptive weights are assigned to each feature, and the weighted features are input into the change vector analysis to obtain the change intensity and direction of the difference image; the expectation–maximisation algorithm is used to threshold the change intensity of the difference image objects to extract the change objects; the $K$ -means algorithm is used to cluster the change direction of the difference image objects to identify and distinguish different change types.

The organisation of this book is divided into 6 chapters (Fig. 1.4). This chapter is the introduction; Chap. 2 is the research area and research data; Chap. 3 is the generation of image objects and the optimal selection of segmentation parameters;

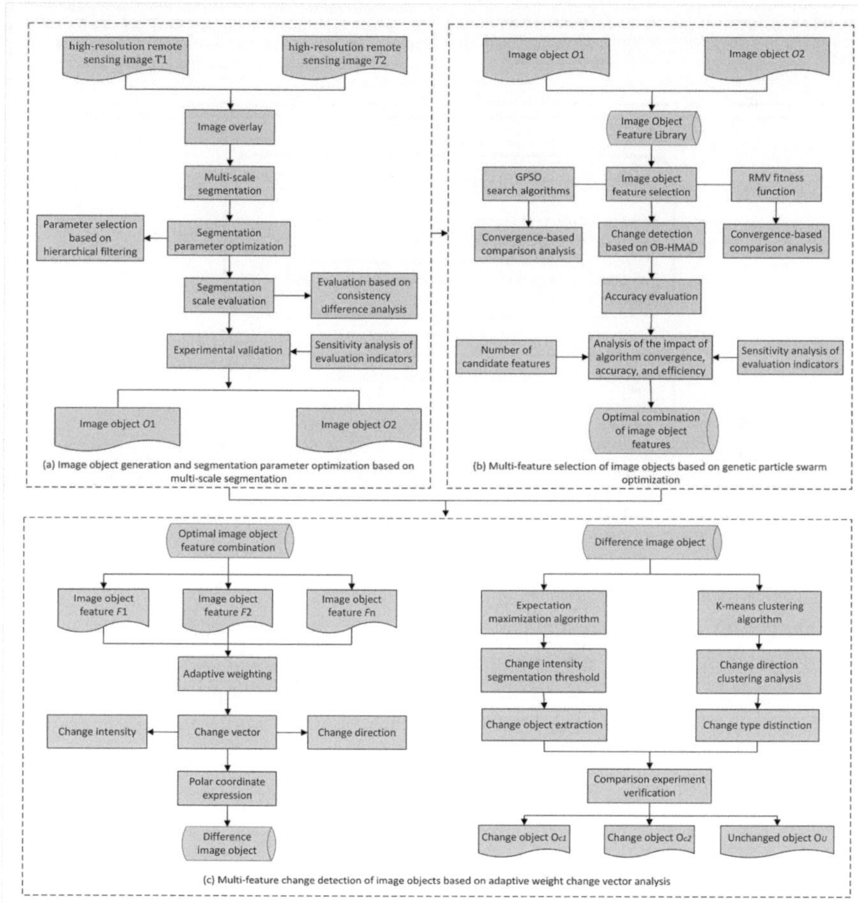


**Fig. 1.3** Technical route diagram of this research

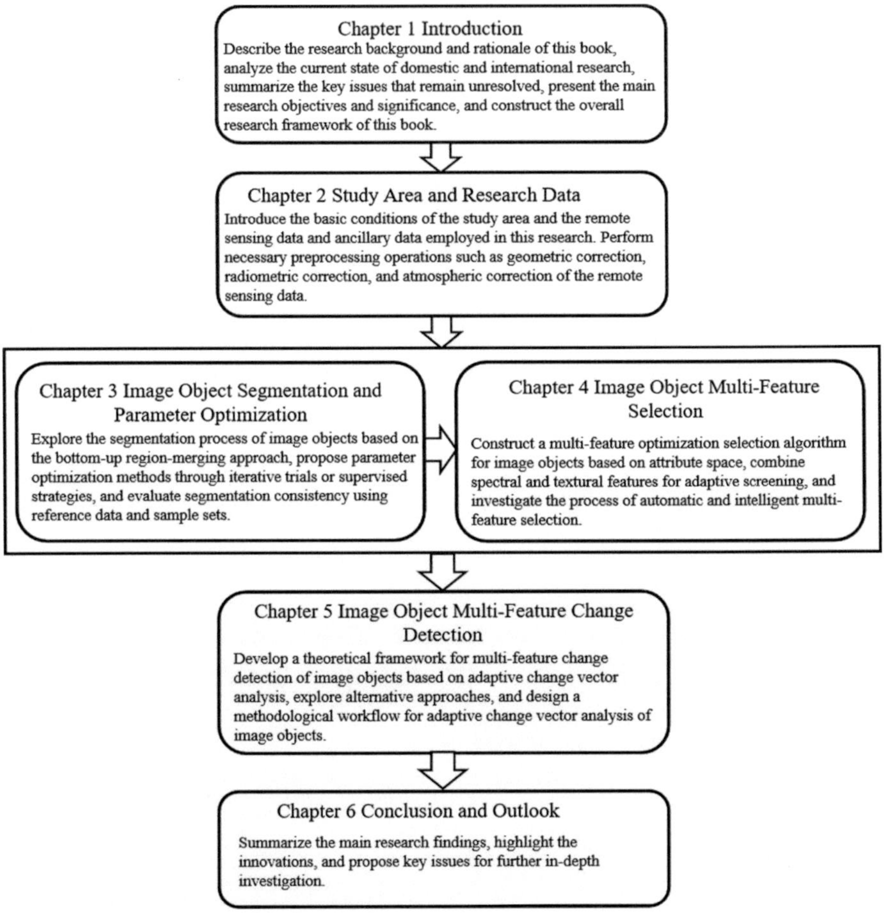


**Fig. 1.4** Chapter arrangement of this book

Chap. 4 is the multi-feature selection of image objects; Chap. 5 is the multi-feature change detection of image objects; Chap. 6 is the conclusion and outlook. Among them, Chaps. 3–5 are the core part of the book, with each chapter progressing and closely linked to the next. The research in Chap. 3 is the premise and guarantee for the latter two, and Chaps. 3 and 4 provide the necessary conditions for the research in Chap. 5. The overall research content of the book always revolves around the main line of "object generation - multi-feature selection - change detection".

# Chapter 2
# Study Area and Research Data

This book selects urban and farmland areas in Beijing as research areas for different types of land cover, using high-resolution images from Worldview-2/3, SPOT-6, and ZY-3 as research data. Reference data is constructed through field surveys and visual judgement, used to evaluate the effectiveness and applicability of the methods in this book. Additionally, this chapter also performs pre-processing operations on the high-resolution remote sensing data, such as geometric correction, orthorectification, relative radiometric correction, and image fusion, to ensure the comparability of the data.

## 2.1 Overview of the Study Area

The study area for this experiment is the Beijing region, with central coordinates: 40° 06′ 10″ N, 116° 25′ 26″ E (Fig. 2.1a). The geographical features of the area are plains, with a dry climate, an average annual temperature of 11.9 °C, and an average annual precipitation of 600–700 mm. The northern area is mainly a mixture of farmland, forest, and low-rise buildings; the southern area is urban, dominated by medium and high-rise buildings and vegetation in parks. To study the changes in different types of objects, two areas are selected as research areas.

Area One is located in and around the Beijing Olympic Park (Fig. 2.1b), spanning the Haidian and Chaoyang districts, with latitude and longitude coordinates ranging from 39° 59′ 47″–40° 00′ 19″ N, 116° 22′ 30″–116° 23′ 54″ E, covering an area of approximately 4 $km^2$. This area was the core area of the 2008 Beijing Olympics, containing various competition venues, urban construction land, artificial grassland, as well as forests and artificial lakes in the Olympic Forest Park. The main surface changes in the area are the construction of competition venues and auxiliary facilities, mainly involving changes in spectral features.

 
Q. Chen et al., *Multiple Feature Object-Based Change Detection Technology for Very High-Resolution Remote Sensing Image*, https://doi.org/10.1007/978-981-95-3917-8_2

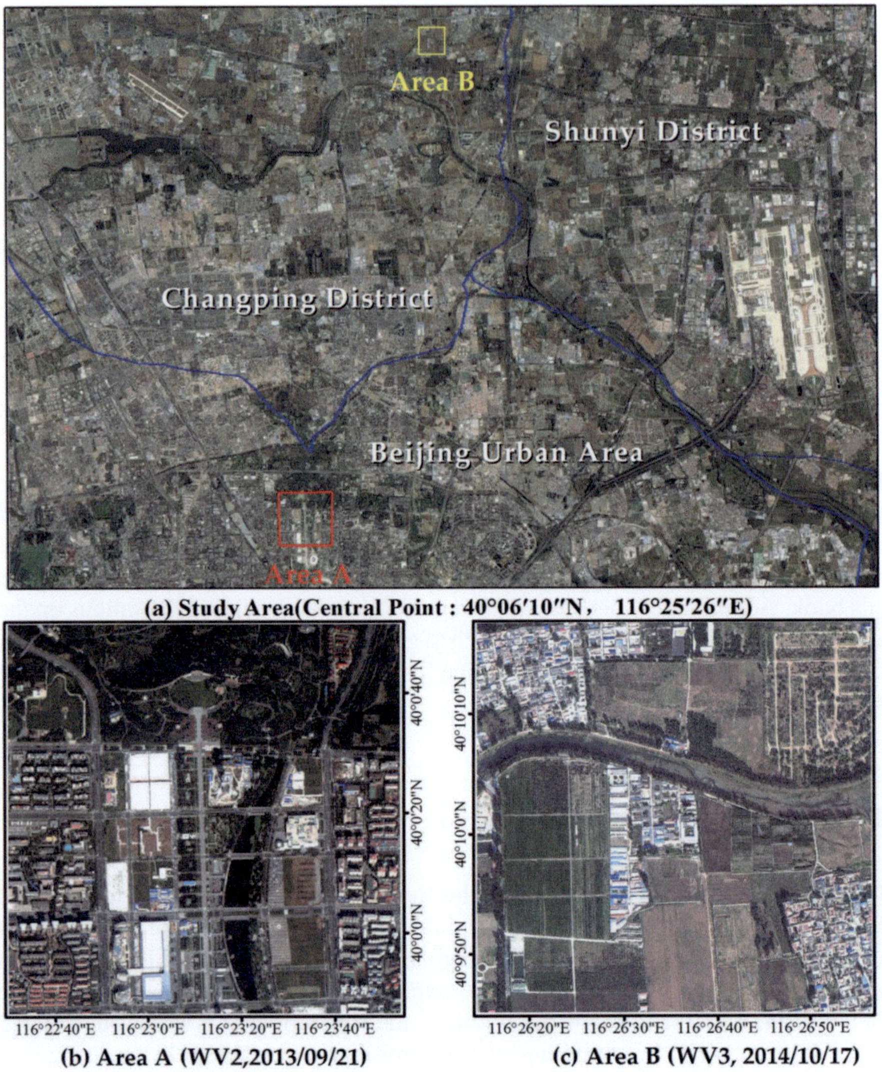


**Fig. 2.1** Schematic diagram of the study area (colour map attached at the end)

Area Two is located in the farmland area around the Xiaotangshan Modern Agricultural Demonstration Garden in Changping District, Beijing (Fig. 2.1c), between Shunsha Road and the North Sixth Ring Road, belonging to the Chaobai River and Shahe River systems. The latitude and longitude coordinates range from 40° 9′ 45″–40° 10′ 19″ N, 116° 22′ 30″–116° 23′ 54″ E, covering an area of approximately 1.5 $km^2$, including farmland, forests, residential houses, and rivers, etc. The main surface changes in the area are the pattern changes of crops planted in the farmland and the natural changes of seasonal alternation, mainly involving changes in texture patterns.

## 2.2 Research Data

### 2.2.1 *High-Resolution Remote Sensing Data*

The study uses high spatial resolution remote sensing data, selecting multispectral high-resolution images from Worldview-2 (WV-2) (0.5 m/2 m), Worldview-3 (WV-3) (0.34 m/1.8 m), SPOT-6 (1.5 m/4 m), and ZY-3 (3.6 m/5.8 m) as research data. Among them, the image types covering Area One are WV-2 and SPOT-6, and the image types covering Area Two are WV-2, WV-3, and ZY-3. Details of the data sources can be found in Table 2.1.

Worldview-2/3 are commercial imaging satellite systems from the American company Digital-globle. WV-2 was launched on October 6, 2009, equipped with a 0.5 m panchromatic and 1.8 m spatial resolution multispectral sensor, featuring four standard spectral bands (red, green, blue, near-infrared), and four additional bands (coastal, yellow, red edge, and near-infrared 2). WV-3 was launched on August 13, 2014, providing a maximum 0.31 m resolution panchromatic and 1.24 m 8-band multispectral imagery.

SPOT-6 is an Earth observation satellite system developed by the French Space Research Centre, launched on September 9, 2012, providing 1.5 m panchromatic and 6 m 4-band multispectral VHR remote sensing data. The ZY-3 mapping satellite is China's first civilian high-resolution mapping satellite, launched on January 9, 2012, equipped with four optical cameras, including a 2.1 m panchromatic CCD camera and a 5.8 m 4-band multispectral camera, as well as two 3.6 m forward and backward viewing panchromatic CCD cameras, which can be used for stereoscopic observation.

**Table 2.1** Details of the research remote sensing data

| Study area | Research data | Spatial resolution/m | Time phase | Number of reference objects | ID |
|---|---|---|---|---|---|
| Area One | WV-2 (Multispectral) | 2 | 2012-09-14 | 26 | WV-2_2012 |
| | WV-2 (Multispectral) | 2 | 2013-09-21 | 26 | WV-2_2013 |
| | SPOT-6 (Panchromatic Fusion) | 1.8 | 2015-08-27 | 22 | FSPOT-6_2015 |
| Area Two | WV-2 (Panchromatic Fusion) | 0.5 | 2010-09-12 | 31 | FWV-2_2010 |
| | WV-3 (Panchromatic Fusion) | 0.34 | 2014-10-17 | 30 | FWV-3_2014 |
| | ZY-3 (Multispectral) | 5.8 | 2015-09-03 | 24 | ZY-3_2015 |

### 2.2.2 *Ground Reference Data*

To evaluate the effectiveness of the methods presented in this book when applied to high-resolution remote sensing data, a set of reference data needs to be constructed as test sample data to assess the reliability and accuracy of these methods. In line with the research objectives of this book, we mainly investigate the changes in various land objects from 2010 to 2015, as well as the main modes of change. The ground reference data in this book is primarily established through on-site field surveys and visual judgement.

On-site field surveys involve visiting the research area in person, identifying the main types of land objects one by one, and obtaining information about the changes in these main land object types through questioning and observation. Visual judgement is based on expert knowledge to determine the main areas and types of change. The reference data constructed in this way is shown in Fig. 2.2.

## 2.3 Data Preprocessing

Errors generated in the acquisition, processing, and analysis of remote sensing image data, along with the instability of the sensor's flight attitude, interference from solar radiation, and other factors causing linear or non-linear distortion, as well as spectral inconsistency, can all cause significant interference in the change detection of multi-temporal remote sensing images (Ma 2006). Therefore, it is necessary to carry out relevant data preprocessing operations on the remote sensing image data to eliminate

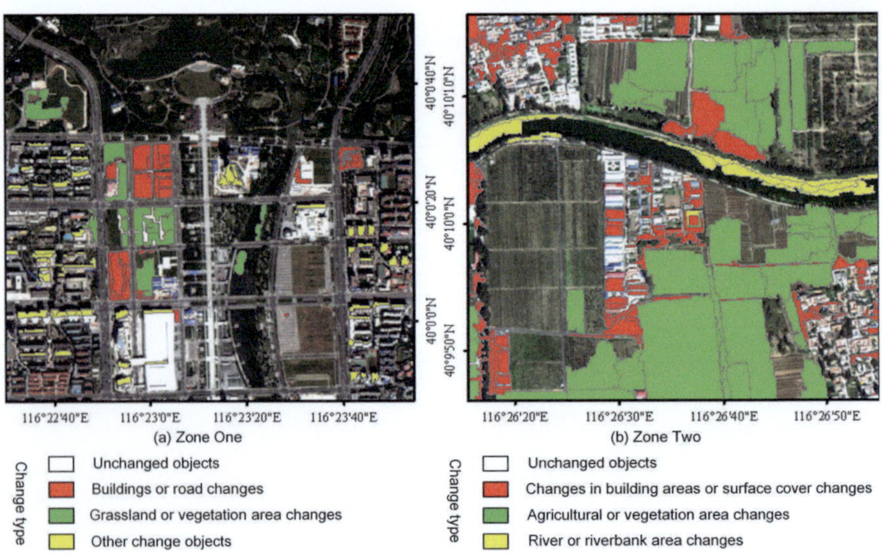


**Fig. 2.2** Ground reference data (colour map attached at the end)

or reduce the impact of these factors, thereby ensuring a certain comparability of multi-temporal remote sensing images and making subsequent image object analysis and change detection possible. Remote sensing image data preprocessing mainly includes geometric correction, orthorectification, radiometric correction, and image enhancement operations.

### 2.3.1 Geometric Correction and Orthorectification

The precise geometric correction of multi-temporal remote sensing images is an important prerequisite and key guarantee for change detection. By overlaying remote sensing images of the same area at different times, the geometric deformation between the reference image and the image to be registered is corrected, geometric registration is carried out, and the comparability between images or image objects is ensured (Coppin et al. 1996; Singh 1989). This book uses a relative geometric registration method based on polynomial transformation, that is, with the help of some ground control points (GCP), the homonymous pixels of multi-temporal remote sensing images are accurately overlapped through the method of polynomial geometric transformation.

The specific process of geometric registration is divided into three steps (Chen 2012): ① Select control points corresponding to the remote sensing image based on the precisely registered remote sensing image and field survey data as reference data; ② Determine the geometric correction model, the commonly used one is the second-order polynomial transformation model; ③ Pixel resampling. This book uses the nearest neighbour pixel method for pixel sampling and controls the correction error (RMSE) within 0.5 pixels.

Orthorectification is the process of orthogonally projecting the original image, using a digital elevation model to correct the original image, correcting the geometric deformation of the image caused by the curvature of the Earth, terrain undulation, and changes in the sensor's exterior orientation elements, and converting it into an orthoimage (Chen 2007). The orthorectification in this book is completed in the Orthorectification module of the ENVI software.

### 2.3.2 Relative Radiometric Correction

Relative radiometric correction can reduce the solar radiation differences between multi-temporal images, reducing the noise interference caused by non-geophysical information changes, ensuring the comparability and consistency of remote sensing images at different times (Liu 2008). Assuming that there is a linear relationship between the pixel grey values of images at different times, we can select pairs of geophysical samples in multi-temporal images where the radiometric values have not changed to establish a linear mapping equation. By solving the parameters in the equation, we can obtain this type of linear relationship (Mei 2001).

Robust regression (RR) is an iterative weighted least squares regression algorithm, where each iteration reassigns weights to the residuals of the previous iteration, assigning different weights to points with different fitting effects. This is a continuous iterative process. RR fits the majority of the data, while ignoring some outliers, thus making the fitting curve less susceptible to the interference of outliers. In change detection research, change information can be approximated as outliers, and relative radiometric correction based on unchanged information can be implemented using robust regression methods.

### 2.3.3 Image Fusion

To enrich the spectral feature information of high-resolution panchromatic images, panchromatic data can be fused with multispectral data to form fused images with high spatial resolution and multi-band spectra. Common image fusion methods include colour technology transformation, arithmetic operation, and image transformation methods. This book uses the Gram-Schmidt spectral sharpening fusion method in image transformation methods. This transformation eliminates redundant information by orthogonally transforming multi-dimensional images. The information of each component of the image will not change or be lost, and it can maximally retain the change information of each band of the image, causing minimal impact on subsequent change detection research. The process of the Gram-Schmidt image fusion method is shown in Fig. 2.3 (Bruzzone et al. 2000a).

All high-resolution remote sensing data are preprocessed, and the experimental data obtained by cropping are shown in Fig. 2.4.

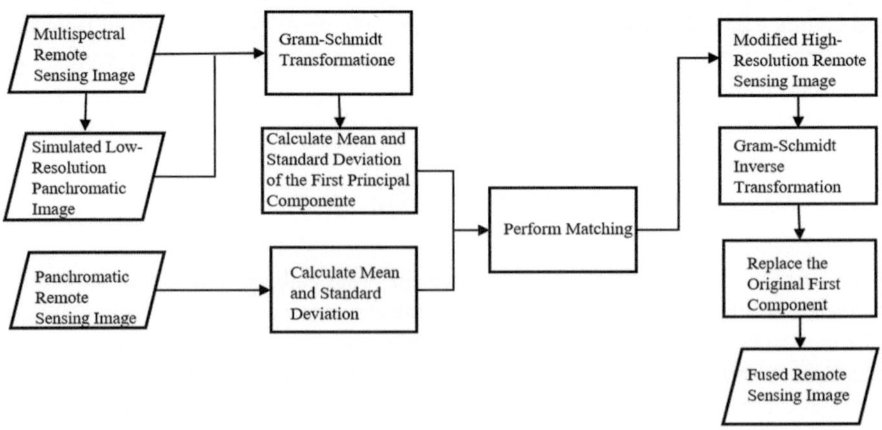


**Fig. 2.3** Process of Gram-Schmidt transformation image fusion

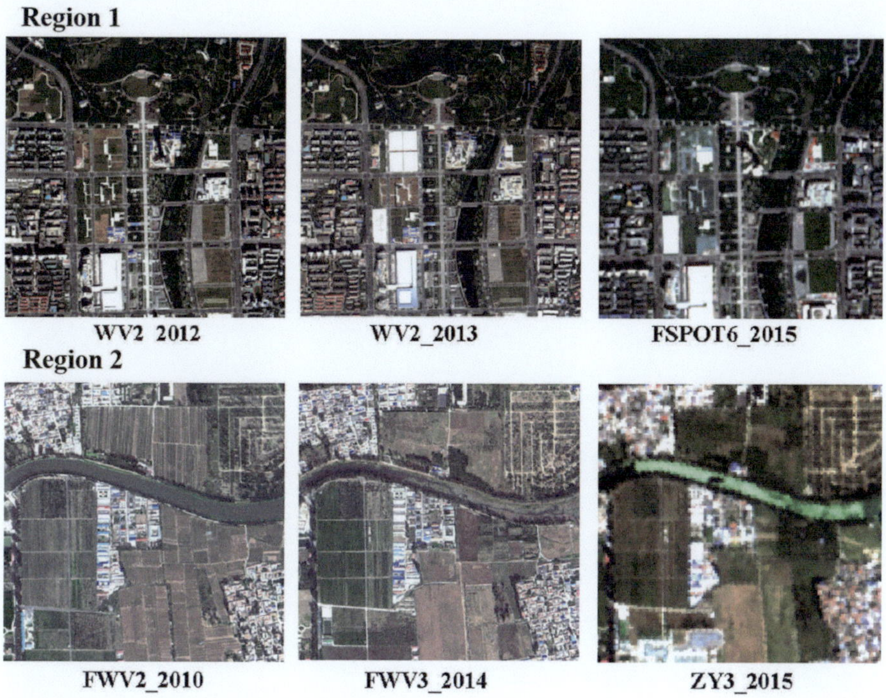


**Fig. 2.4** Preprocessed experimental data

# Chapter 3
# Image Object Generation and Optimal Segmentation Parameter Selection

This chapter introduces the image object generation process based on the multi-scale segmentation algorithm and conducts research on the optimal selection of parameters such as segmentation scale. Based on multi-scale segmentation, the multi-band segmentation quality index meanNSQI is constructed based on the internal homogeneity and external heterogeneity of the image object, and the optimal segmentation parameters are automatically selected in a hierarchical screening manner. Then, the geometric/quantity consistency difference between the segmented object and the reference object is used as the evaluation criterion to evaluate the optimal parameters. The reliability of the segmentation parameter optimization method in this book is verified using high-resolution remote sensing images of different resolutions as experimental data. In addition, a sensitivity analysis of the consistency difference evaluation index was conducted to explore the impact of multi-temporal image overlay on the segmentation parameter optimization results.

The generation of high-resolution remote sensing image objects mainly depends on image segmentation technology. Image segmentation is the basis of object-oriented image analysis methods. Image segmentation is a technology and process that divides images into regions with specific characteristics and extracts objects of interest. Through segmentation, the target objects contained in the image and the features describing the object attributes can be obtained, which is more conducive to subsequent image analysis and understanding. Image objects with reasonable size and matching boundaries are the prerequisite and guarantee for subsequent research, so how to generate reasonable and accurate image objects using image segmentation is a major research issue.

According to the content of Sect. 1.2.1, this book mainly uses the multi-scale segmentation algorithm as the main means of generating image objects, where the key issue is the optimization and evaluation of algorithm segmentation parameters. Under the principle of maximizing the homogeneity within objects and the heterogeneity between objects, a segmentation quality evaluation index is proposed. For the top-down segmentation characteristics in multi-scale segmentation technology,



Q. Chen et al., *Multiple Feature Object-Based Change Detection Technology for Very High-Resolution Remote Sensing Image*, https://doi.org/10.1007/978-981-95-3917-8_3

with the spectral/shape heterogeneity of the object as the weight factor, the spectral concentration of internal pixels is calculated by the standard deviation of the spectral mean of the object pixels, and the correlation or heterogeneity of external objects is calculated by the Moran index. By maximizing the internal homogeneity and external heterogeneity of the segmented object features, the segmentation quality evaluation index (NSQI) is constructed; based on meanNSQI (the mean of NSQI in each band), the optimal segmentation parameters of remote sensing images of different resolutions are obtained through a hierarchical screening method.

Consistency difference analysis is used as the evaluation method to verify the reliability of the optimal segmentation parameters. A sensitivity analysis of the consistency difference evaluation index and the segmentation scale parameter is conducted to determine its reliability as a segmentation object quality evaluation index; and the consistency issue of overlay images and original single-frame images is discussed, analysing the impact of image overlay in OBCD on image segmentation.

This chapter is divided into six sections. Section 3.1 discusses the theory of multi-scale image segmentation; Sect. 3.2 covers the construction of meanNSQI and the optimisation process of segmentation parameters based on hierarchical screening; Sect. 3.3 analyses the geometric/quantitative consistency differences between reference objects and segmentation objects, and evaluates the segmentation objects; Sect. 3.4 presents experiments and result analysis; Sect. 3.5 explores the sensitivity analysis of consistency difference evaluation indicators, as well as the analysis of consistency issues in multi-temporal image segmentation objects; Sect. 3.6 summarises the content of this chapter (Fig. 3.1).

## 3.1 Multi-scale Image Segmentation Theory

In the method based on region merging, the multi-scale segmentation algorithm provided by the eCognition Developer platform, which fully utilises object features and inter-class relationships, is widely accepted in the industry. This book selects the multi-scale segmentation algorithm as the method for generating image objects, and analyses and optimises its parameters, aiming to make the segmentation objects more suitable for subsequent image analysis and change detection.

The basic idea of the multi-scale segmentation algorithm (Baatz et al. 2000; Chen et al. 2006; Witharana et al. 2014) is to calculate the feature values of spectral feature heterogeneity and shape feature heterogeneity of each band of the image, and according to the weight of each band, calculate the weighted value of all bands of the image. By repeating iterative calculations, until the comprehensive weighted value of the spectrum and shape of all objects is greater than a certain specified threshold, the multi-scale segmentation of the image is completed. For image band $i$, its heterogeneity can be obtained from the following formula:

$$f_i = w_i h_{\text{shape}} + (1 - w_i) h_{\text{color}} \tag{3.1}$$

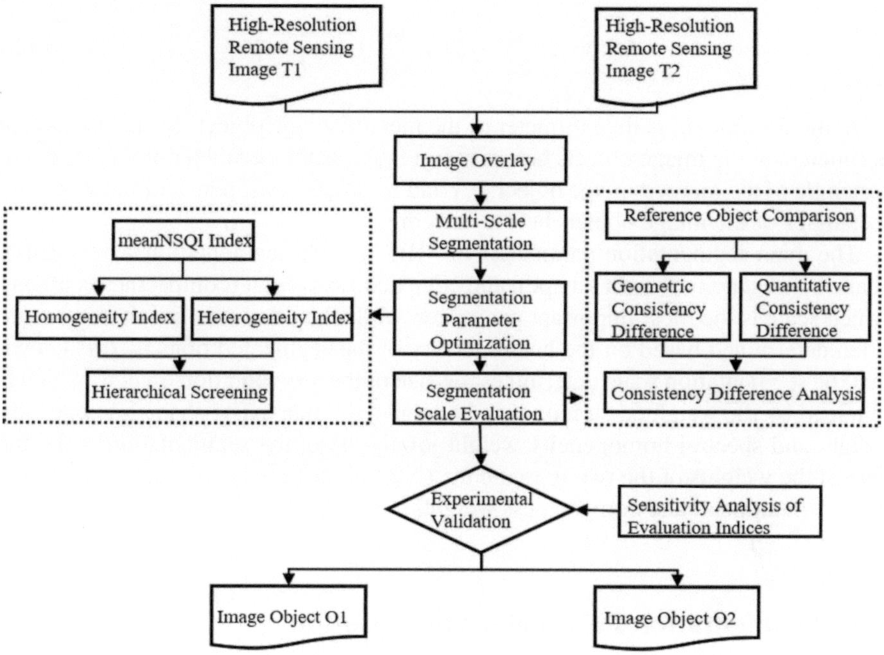


**Fig. 3.1** Process of image object generation and segmentation parameter optimisation

In the formula, $w_i$ is the weight of the spectral heterogeneity $h_{\text{color}}$ of band $i$, $0 \leq w_i \leq 1$, set by the user. $h_{\text{color}}$ The calculation formula is as follows:

$$h_{color} = \sum_{c} w_c(n_m\sigma_{c,m} - (n_{o1}\sigma_{c,o1} + n_{o2}\sigma_{c,o2})) \tag{3.2}$$

In the formula, $w_c$ is the weight of the band to be segmented; $c$ is the number of image bands; $n_m$ is the number of pixels in the merged object; $\sigma_{c,m}$ is the standard deviation of the merged image; $n_{o1}$ and $n_{o2}$ are the number of pixels in the object before merging; $\sigma_{c,o1}$ and $\sigma_{c,o2}$ are the standard deviations of the image before merging.

Shape heterogeneity $h_{\text{shape}}$ is determined by shape compactness $h_{\text{cp}}$ and shape smoothness $h_{\text{sm}}$:

$$h_{\text{shape}} = w_{\text{cp}}h_{\text{cp}} + \left(1 - w_{\text{cp}}\right)h_{\text{sm}} \tag{3.3}$$

In the formula, $w_{\text{cp}}$ is the weight of shape compactness, and the calculation formulas for shape compactness and shape smoothness are

$$h_{\text{cp}} = n_m\frac{l_m}{\sqrt{n_m}} - \left(n_{o1}\frac{l_{o1}}{\sqrt{n_{o1}}} + n_{o2}\frac{l_{o2}}{\sqrt{n_{o2}}}\right) \tag{3.4}$$

$$h_{\text{sm}} = n_m \frac{l_m}{b_m} - \left( n_{o1} \frac{l_{o1}}{b_{o1}} + n_{o2} \frac{l_{o2}}{b_{o2}} \right) \tag{3.5}$$

In the formula, $l_m$ is the perimeter of the merged image object; $l_{o1}$ and $l_{o2}$ are the perimeters of the image objects before merging; $b_m$ is the perimeter of the bounding rectangle of the merged image object; $b_{o1}$ and $b_{o2}$ are the perimeters of the bounding rectangles of the image objects before merging.

The main segmentation parameters of MRS include segmentation scale (scale), shape homogeneity weight (shape), and compactness weight (compactness), among which scale is the most important parameter, while the other two weight factors are often determined based on the heterogeneity of the segmented objects. It is known that the segmentation scale determines the size of the segmentation objects, which is obtained by the weighted sum of each homogeneity index. The shape homogeneity weight and spectral homogeneity weight jointly affect the segmentation scale, the sum of the weights of the two is 1, and the shape homogeneity weight is determined by the smoothness and compactness weights.

## 3.2 Segmentation Parameter Optimisation

The ideal image segmentation results have the following characteristics: ① The segmented objects have good homogeneity internally, that is, there are fewer mixed objects and the concentration within the object is high; ② The segmented objects have good heterogeneity with other surrounding adjacent objects, that is, the separability between objects is high. Therefore, based on relevant statistical principles, a Normalised Segmentation Quality Index (NSQI) can be constructed to characterise the quality of segmentation results and to select the optimal segmentation parameters.

### *3.2.1 Homogeneity Index*

The quality parameters of the segmented object features need to be evaluated. In object-oriented segmentation, a smaller segmentation scale can make the change information too fragmented, causing the object to lose its meaning consistent with the real geographical object, while a larger scale can mask the change information at a smaller scale. It is generally believed that larger objects can reduce the pixel differences within the image object. If it is assumed that smaller objects have more internal differences, it is beneficial to enhance the spectral concentration of the segmented objects (Edwards and Cavalli-Sforza 1965). According to statistical theory, the standard deviation within the object is used to characterise the feature concentration of the segmented objects at different scales, replacing the pixel spectral value with the object spectral mean, and normalising it with the standard deviation of the pixel spectral value, which is a value between 0 and 1. Generally, the larger the value of $C$, the higher the spectral concentration of the segmented object, indicating that the object has good internal homogeneity.

$$C = \sqrt{\frac{\sum_{i=1}^{n} m_i (\overline{x} - \overline{x}_i)^2}{\sum_{j=1}^{m} (\overline{x} - x_j)^2}} \tag{3.6}$$

In the formula, $n$ is the number of objects; $m$ is the number of pixels; $m_i$ is the number of pixels contained in the $i$th object; $\overline{x}$ is the spectral mean of a certain band of the image; $\overline{x}_i$ is the spectral mean of this band of the $i$th object; $x_j$ is the spectral value of this band of the $j$th pixel.

### 3.2.2 Heterogeneity Index

The segmented object not only requires a high spectral concentration internally, but also requires a high separability between objects. The heterogeneity between objects is mainly reflected through the correlation between objects, that is, the higher the correlation, the worse the separability. This book uses the Moran index to express the heterogeneity between objects. The Moran index is an indicator representing the spatial correlation of objects (Shortridge 2007), and its formula is

$$I = \frac{n \sum_{i=1}^{n} \sum_{j=1}^{n} w_{ij} (y_i - \overline{y}) (y_j - \overline{y})}{S_0 \left( \sum_{i=1}^{n} (y_i - \overline{y})^2 \right)} \tag{3.7}$$

$$S_0 = \sum_{i=1}^{n} \sum_{j=1}^{m} w_{ij} \tag{3.8}$$

In the formula, $n$ is the total number of objects; $w_{ij}$ is the adjacency relationship between objects $i$ and $j$. If objects $i$ and $j$ are adjacent, then $w_{ij} = 1$; if they are not adjacent, then $w_{ij} = 0$; $\overline{y}$ is the spectral mean of the entire image; $y_i$ is the spectral mean of the $i$th object; The lower the $I$, the lower the correlation between the two objects, and the higher the spectral separability.

### 3.2.3 Segmentation Quality Evaluation Index

An ideal segmented object should have a high internal spectral concentration and spectral separability between external objects. Therefore, the Normalised Segmentation Quality Index (NSQI) should satisfy the following equation:

$$\mathrm{NSQI} = \left| \frac{C - I}{C + I} \right| \tag{3.9}$$

NSQI is calculated for each band. To make full use of the information of multiple bands, this book calculates the NSQI for each band and obtains the meanNSQI based on the research of Drăguţ et al. (2014):

$$\mathrm{meanNSQI} = (\mathrm{NSQI}_1 + \mathrm{NSQI}_2 + \cdots + \mathrm{NSQI}_n)/n \tag{3.10}$$

In the formula, $n$ is the number of bands. To ensure the consistency of image objects at different times in the subsequent change detection process, the same parameter settings are adopted for each band, and all participate in the MRS algorithm segmentation. The weight of each band is set to 1.

For any segmentation scale $S_i$, the segmentation object is influenced by the shape heterogeneity weight $w_{\mathrm{shape}}$ and the compactness weight $w_{\mathrm{cp}}$. If the set of segmentation objects is defined as $F(S, w_{\mathrm{sh}}, w_{\mathrm{cp}})$, for a certain segmentation scale $S$, there exists a maximum average segmentation quality index $\mathrm{NSQI}_{\max}$ $(S)$.

$$\mathrm{meanNSQImax}(\mathrm{meanNSQI}_{\mathrm{sh,cp\,max}} \tag{3.11}$$

When the average segmentation quality index NSQI $_{\max}$ reaches its maximum value, the corresponding segmentation scale $S$ is the optimal scale, and its corresponding set of parameters $(S, w_{\mathrm{shape}}, w_{\mathrm{cp}})$ are the optimal segmentation parameters.

### 3.2.4 *Optimal Segmentation Scale Selection Based on Hierarchical Filtering*

Based on the statistical analysis of the reference object and the corresponding object, the main focus is on the changes in NSQI under different segmentation parameter conditions, thereby measuring the changes in geometric and quantity consistency differences. The segmentation parameters corresponding to the minimum NSQI are selected as the optimal segmentation scale parameters.

Due to the unknown nature of the alternative segmentation scale interval, and to avoid the interference of a large number of homogeneous objects or pixels at a small scale, a hierarchical filtering method is used to calculate and compare the meanNSQI index. First, with [1, 1000] as the scale interval and 100 as the step size, calculate the meanNSQI index at each scale and select the maximum value corresponding to the segmentation scale as the alternative scale $S$; with $[S - 50, S + 50]$ as the scale interval and 10 as the step size, calculate the meanNSQI index at each scale and select the maximum value corresponding to the segmentation scale to update the alternative scale $S$; with $[S - 5, S + 5]$ as the scale interval and 1 as the step size, calculate the meanNSQI index at each scale and select the maximum value corresponding to the segmentation scale as the optimal scale $S$. The specific process is shown in Fig. 3.2.

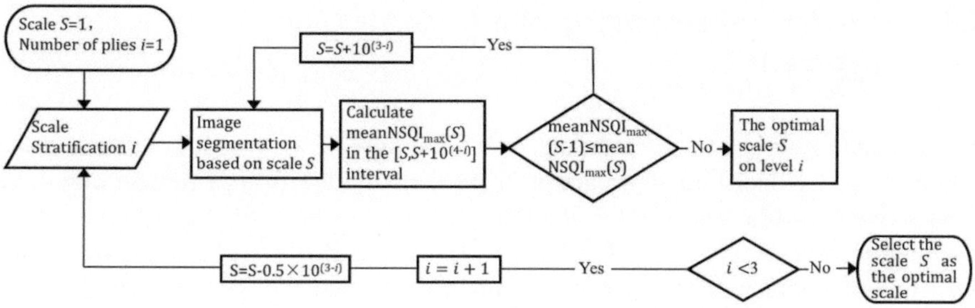


**Fig. 3.2** Automatic scale selection process of the multi-scale segmentation algorithm

Depending on the image resolution, select the appropriate segmentation scale range and step size as the independent variable for statistical analysis. For the corresponding shape heterogeneity weight and smoothness weight, set the variation interval [0.1, 0.9] and step size 0.1 as the sub-items of this segmentation scale, then each segmentation scale will have $9 \times 9 = 81$ sub-items. Therefore, if the variation interval of the segmentation scale is [10, 100] and the step size is 10, it will form a scatter plot containing 810 points, where the x-axis is the segmentation scale containing various weights, each column has 81 points; the y-axis is the corresponding evaluation index. At the same time, construct a box error graph to analyse the expectation and variance of the scatter data set under each segmentation scale.

## 3.3 Evaluation of Optimal Segmentation Parameter Selection

The consistency difference analysis is based on the corresponding relationship between the segmentation object and the reference object. It focuses on the difference in consistency between the segmentation object and the real object. By analysing the geometric or quantity correspondence between the two, a series of evaluation indicators are constructed to evaluate the segmentation quality of the image object. Therefore, this empirical model mainly includes three parts: ① difference analysis of geometric or quantity consistency corresponding to the reference object and the segmentation object; ② a series of evaluation standards are constructed based on this relationship analysis; ③ statistical analysis of the correlation between various evaluation indicators and the segmentation scale, to select the optimal segmentation scale.

### 3.3.1 Analysis of Geometric and Quantity Consistency Differences

A reasonable reference object should be a geographical area with boundaries, and it is a basic regional unit that cannot be further divided from a human cognitive perspective (Castilla et al. 2007; Lang et al. 2010), it can be easily identified from the surrounding area. According to the theoretical framework of OBIA and MRS, various basic regional units with spatial domain and intrinsic connections can be merged into a new composite region. This composite region can also be further merged into larger regional objects based on its internal homogeneity and domain area. This is the image segmentation process based on regional merging. The reference object of interest in this book is the initial basic regional unit, which is artificially selected from the image based on field investigation. Therefore, it may only be a small part of the experimental area and must be included within the experimental area.

In the image segmentation process, the entire image is divided into a series of discrete, adjacent, non-intersecting segmentation objects. Theoretically, the smallest segmentation object can be a single pixel, and the largest can be the entire image. For the same area, the number of objects segmented under different scales is also different. Therefore, in order to better analyse the geometric relationship between the reference object and the corresponding segmentation object, it is necessary to define the corresponding segmentation object: ① All sub-objects belonging to the segmentation object dataset can be candidate objects; ② When the candidate object intersects with the reference object, the candidate object is the corresponding segmentation object; ③ If the area of intersection between the candidate object and the reference object exceeds half of the area of the candidate object or the reference object, this candidate object is called a perfectly corresponding object. This judgement standard is the areal-overlap criterion (Clinton et al. 2010). Figure 3.3 describes the actual situation of various image objects.

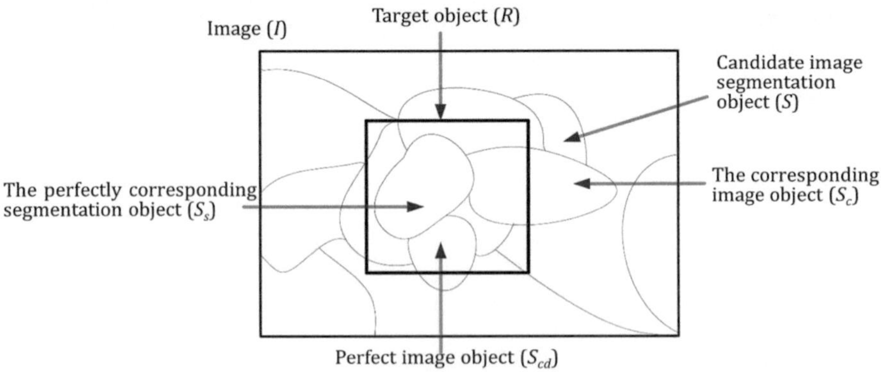


**Fig. 3.3** Schematic diagram of various image objects

If $r_i$ represents the $i$th component or sub-object in the reference object $R$, which is a polygon contained in $R$, and the reference object is composed of these polygons, $R = \{r_i : i = 1, 2, \ldots, m\}$ $R = \{r_i: i = 1, 2, \ldots, m\}$; $s_j$ represents the $j$th object in the candidate segmentation object set $S$, which is a separate segmentation object, $S = \left\{s_j : j = 1, 2, \ldots, n\right\}$. Then the corresponding segmentation object dataset $S_c = \{s_{ck} : k = 1, 2, \ldots, l\}$, it is known that $S_c$ is a subset of $S$, and $l \leq n$. When $S_c$ meets the areal-overlap criterion, it is the appropriate corresponding object $S_s = \{s_{su} : u = 1, 2, \ldots, v\}$, it is known that $S_s$ is a subset of $S_c$, and $v \leq l$, it is known:

$$\frac{|\mathrm{S_{su}} \cap \mathrm{r_i}|}{\mathrm{r_i}} \geq 0.5 \text{ or } \frac{|\mathrm{S_{su}} \cap \mathrm{r_i}|}{\mathrm{S_{su}}} \geq 0.5 \tag{3.12}$$

Generally speaking, there is inevitably some difference between the reference object and the corresponding object, which can be divided into geometric differences and quantity differences. Geometric differences mainly refer to the inconsistency in the geometric overlay relationship between the reference object and the corresponding object. This geometric overlay relationship is divided into three types: overlap, over-segmentation, and under-segmentation, as shown in Fig. 3.4. If the intersection area of the reference object and the corresponding object is the overlap area; if part of the reference object falls outside the corresponding object, then this part of the area is the over-segmentation area; if part of the corresponding object area is not covered by the reference object, then this part of the area is the under-segmentation area (Marpu et al. 2010). According to the areal-overlap criterion, the over-segmentation and under-segmentation areas should be smaller than the overlap area. At the same time, there is also a quantity difference between the reference object and the corresponding object, that is, the inconsistency of the corresponding quantity. Since the corresponding object can be as small as a single pixel, considering only the geometric consistency difference cannot guarantee the consistency of segmentation. As shown in Fig. 3.4a, a reference object corresponds to three corresponding objects, which is a one-to-many difference; Fig. 3.4b shows a reference object exactly corresponding to a corresponding object, which is a one-to-one difference; Fig. 3.4c shows multiple reference objects corresponding to a corresponding object, which is a many-to-one difference. These geometric and algorithmic differences together describe the consistency difference between the reference object and the corresponding object.

### 3.3.2 Construction of Evaluation Indices

Based on the analysis of the geometric and quantity relationship between the reference object and the corresponding segmentation object, the corresponding evaluation index is constructed to describe this geometric and quantity consistency difference.

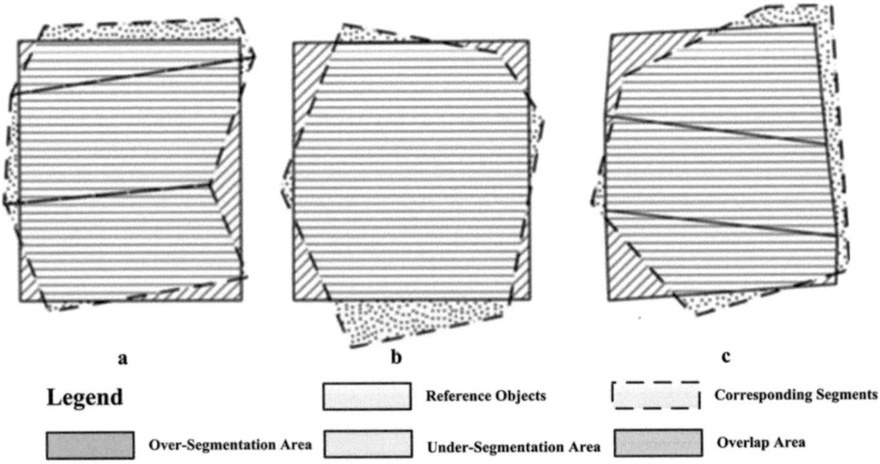


**Fig. 3.4** Schematic diagram of geometric and quantity differences between reference objects and corresponding objects (Liu et al. 2012)

For the geometric consistency difference between the reference object and the corresponding object, the QR (quality indices such as QR (Quality Ratio), OR (Over-segmentation Ratio), and UR (Under-segmentation Ratio), and optimise OR and UR using Euclidean distance to obtain the ED1 (Euclidean Distance 1) index. Their mathematical expressions are as follows:

$$\mathrm{QR} = 1 - \frac{\sum |r_i \cap s_{cu}|}{\sum |r_i \cup s_{cu}|} \tag{3.13}$$

$$\mathrm{OR} = 1 - \frac{\sum |r_i \cap s_{su}|}{\sum |r_i|} = \frac{\sum |r_i - s_{su}|}{\sum |r_i|} \tag{3.14}$$

$$\mathrm{UR} = 1 - \frac{\sum |r_i \cap s_{su}|}{\sum |s_{su}|} = \frac{\sum |s_{su} - r_i|}{\sum |s_{su}|} \tag{3.15}$$

$$\mathrm{ED1} = \sqrt{\frac{\mathrm{OR}^2 + \mathrm{UR}^2}{2}} \tag{3.16}$$

where, $r_i$ is a subset of the reference object dataset $R$; $s_{su}$ is a subset of the corresponding object dataset $S_s$; $|r_i \cap s_{su}|$, $|r_i - s_{su}|$ and $|s_{su} - r_i|$ represent the overlapping, over-segmented, and under-segmented areas respectively; and $|r_i \cup s_{su}|$ represents the combined area of these three regions. QR, OR, and UR all take values in the range [0, 1], with smaller values indicating a better correspondence between the reference object and the corresponding object. QR $= 0$ indicates that the area has neither over-segmentation nor under-segmentation, and the corresponding object perfectly matches the reference object. OR $= 0$ and UR $= 0$ indicate that the area has no over-segmentation and under-segmentation respectively. ED1 measures the distance to the origin in the OR-UR Euclidean space, representing the impact of over-segmentation and under-segmentation. Since both over-segmentation and under-segmentation affect the quality of segmentation, ED1 can be used as an index to evaluate the segmentation scale. When ED1 = 0, it indicates that there is no over-segmentation and under-segmentation, and the geometric consistency difference is minimal, representing the optimal segmentation situation.

These four parameters all represent the geometric consistency difference between the reference object and the corresponding object. As can be seen from the above analysis, there is also a numerical consistency difference between the reference object and the corresponding object. The PSE (Potential Segmentation Error), NSR (Number of Segments Ratio), and ED2 (Euclidean Distance 2) indices are constructed to describe the numerical consistency difference between the reference object and the corresponding object.

$$\mathrm{PSE} = \frac{\sum |s_{su} - r_i|}{\sum |r_i|} \tag{3.17}$$

$$\mathrm{NSR} = \frac{\mathrm{abs}(m - v)}{m} \tag{3.18}$$

$$\mathrm{ED2} = \sqrt{\mathrm{PSE}^2 + \mathrm{NSR}^2} \tag{3.19}$$

In the formula, PSE measures the proportion of the under-segmented area in the reference object, similar to the UR parameter, the smaller the value, the less the under-segmented area; NSR measures the difference between the number of polygons that make up the reference object and the corresponding object as a proportion of the number of polygons in the reference object. When NSR = 0, it indicates that the reference object and the corresponding object have a one-to-one relationship, and the larger the NSR, the more obvious the one-to-many relationship between the reference object and the corresponding object (fewer reference object polygons and more corresponding objects). When the corresponding objects are all single pixels, NSR reaches its maximum value. NSR reflects the numerical consistency difference in the case of over-segmentation; ED2 measures the distance to the origin in the PSE-NSR Euclidean space, expressing the combined effect of PSE and NSR. When ED2 = 0, it indicates that the numerical consistency difference is minimal, representing the optimal segmentation situation. The larger the ED2, the greater the numerical consistency difference. Therefore, ED2 can be used as an index to evaluate the numerical consistency difference between the reference object and the corresponding object.

## 3.4 Experiment and Result Analysis

To address issues such as the optimisation of high-resolution image object generation and segmentation parameters, the MRS algorithm is used as a basis to conduct segmentation experiments on remote sensing images of different spatial resolutions in two experimental areas. This is to verify the effectiveness and applicability of the proposed meanNSQI index in the process of multi-scale segmentation parameter optimisation. Figure 3.5 shows the key steps in the experimental analysis. Firstly, multi-temporal high-resolution remote sensing images are superimposed and segmented as a whole; secondly, the segmentation scale, shape heterogeneity weight, and compactness heterogeneity weight are assigned at different levels, and multi-scale segmentation is carried out based on these segmentation parameters; thirdly, the meanNSQI index at different levels is calculated, and layered screening is carried out to obtain the optimal segmentation parameter group; fourthly, by comparing the reference object with the segmented object, geometric/numerical consistency difference analysis is carried out to obtain various evaluation indices; finally, statistical analysis is carried out to evaluate the quality of segmentation.

### 3.4.1 Experimental Data

To analyse the segmentation of high-resolution remote sensing images at different spatial resolutions, four types of multispectral high-resolution images are selected as research objects: Worldview-2 (0.5 m/2 m), Worldview-3 (0.34 m/1.8 m), SPOT-6 (1.5 m/4 m), and ZY-3 (3.6 m/5.8 m). The experimental area includes two regions,

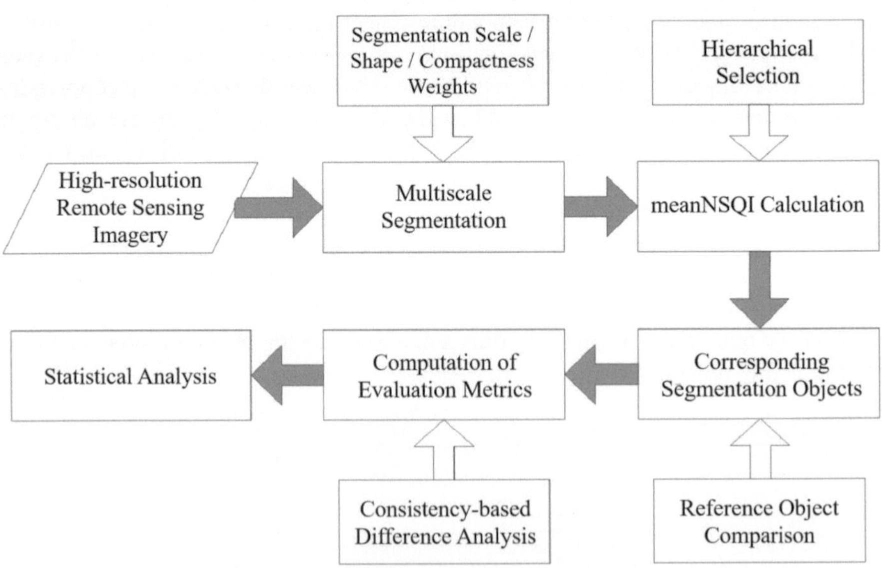


**Fig. 3.5** Key steps in multi-scale image segmentation and parameter optimisation analysis

with the images corresponding to region one being WV-2_2012, WV-2_2013, and FSPOT-6_2015; the images covering region two are FWV-2_2010, FWV-3_2014, and ZY-3_2015.

To explore the applicability of the segmentation quality index-based scale selection method at different spatial resolutions, images of different resolutions were selected for each region. Simultaneously, considering subsequent change detection applications, images of the same/similar resolution were selected for analysis in each region. The reference objects for region one were mainly buildings and grassland areas, as well as some water bodies; the reference objects for region two were mainly agricultural land and forest land.

### *3.4.2 Scale Selection Based on meanNSQI*

According to the descriptions in Sect. 3.2 and Fig. 3.2, multi-scale segmentation was performed on the six images of the two experimental regions, and level filtering was carried out based on meanNSQI. This resulted in the distribution and boxplot of meanNSQI at each level (Figs. 3.6, 3.7, 3.8, 3.9, 3.10 and 3.11). There are 81 parameter combinations under each segmentation scale (9 shape weight parameters and 9 compactness weights, when the scale is 10, its parameter combination is scale10). The distribution of meanNSQI under each parameter combination can be intuitively seen in the distribution diagram, while the boxplot shows the maximum, minimum, mean, and standard deviation of meanNSQI under different parameter combinations at each scale. The specific statistical results are shown in Table 3.1.

Figure 3.6 describes the optimal scale selection situation for the WV-2_2012 image, based on three levels and meanNSQI. At level 1, the scale selection interval is [1:801], with a step size of 100. There is a significant difference in the distribution of meanNSQI under different segmentation scales. The meanNSQI of scale201 is significantly higher than other scales, and the standard deviation is also lower, indicating that at level 1, when the segmentation scale is 201, the segmentation effect is better. According to the description in Sect. 3.3.4, the scale selection interval for level 2 is [151:251], with a step size of 10 for extended analysis. Similarly, scale171 is superior to other scale parameter combinations; the interval for level 3 is [166:176], with a step size of 1. The meanNSQI of each segmentation scale is quite close, with the better parameter combination being scale173.

Figure 3.7 describes the optimal scale selection situation for the WV-2_2013 image, based on three levels and meanNSQI. Similar to the WV-2_2012 image, during the multi-scale segmentation scale selection process for the WV-2_2013 image, there is a significant difference in meanNSQI at each scale at levels 1 and 2, while the mean and standard deviation of meanNSQI at each scale at level 3 are quite close. The most suitable scale is scale168. A direct comparison with Fig. 3.6 shows that the scale selection results for these two images are quite similar, with the preferred scales at levels 1 and 2 being scale201 and scale171, respectively. The preferred scale at level 3 is also very close (scale173 and scale168), indicating that the objects in these two images have a certain similarity and consistency, and it is possible to segment them together.

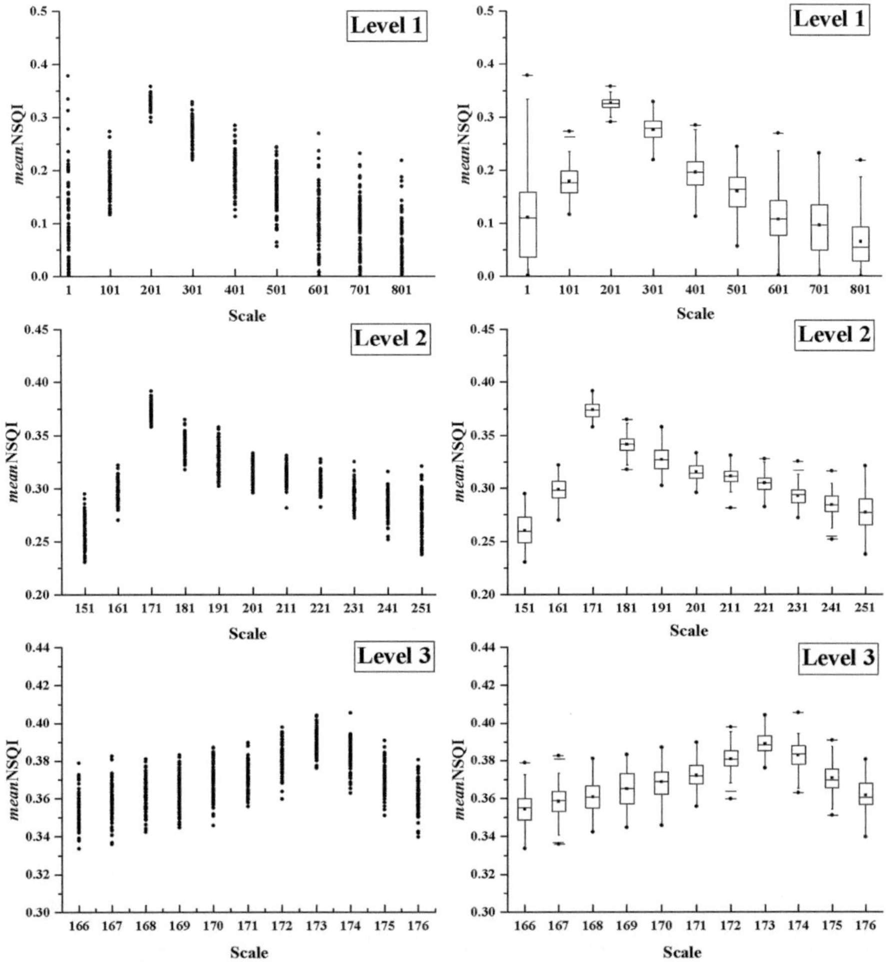


**Fig. 3.6** Distribution and boxplot of meanNSQI at different scales for each level of the WV-2_2012 image

Figure 3.8 describes the optimal scale selection situation for the FSPOT-6_2015 image, based on three levels and meanNSQI. It can be seen that there is a significant difference in meanNSQI at each scale at all levels, with large variations. The optimal scale at levels 2 and 3 is scale211, indicating that compared to other scales, the segmentation effect of scale211 is significantly better.

Figure 3.9 describes the optimal scale selection situation for the FWV-2_2010 image, based on three levels and meanNSQI. At level 1, the meanNSQI of scale201 is higher than other scales, and similarly at level 2, scale211 is superior to other scale parameter combinations. The meanNSQI of each segmentation scale at level 3 is quite close, and the scale208 parameter combination can achieve better segmentation results on the FWV-2_2010 image.

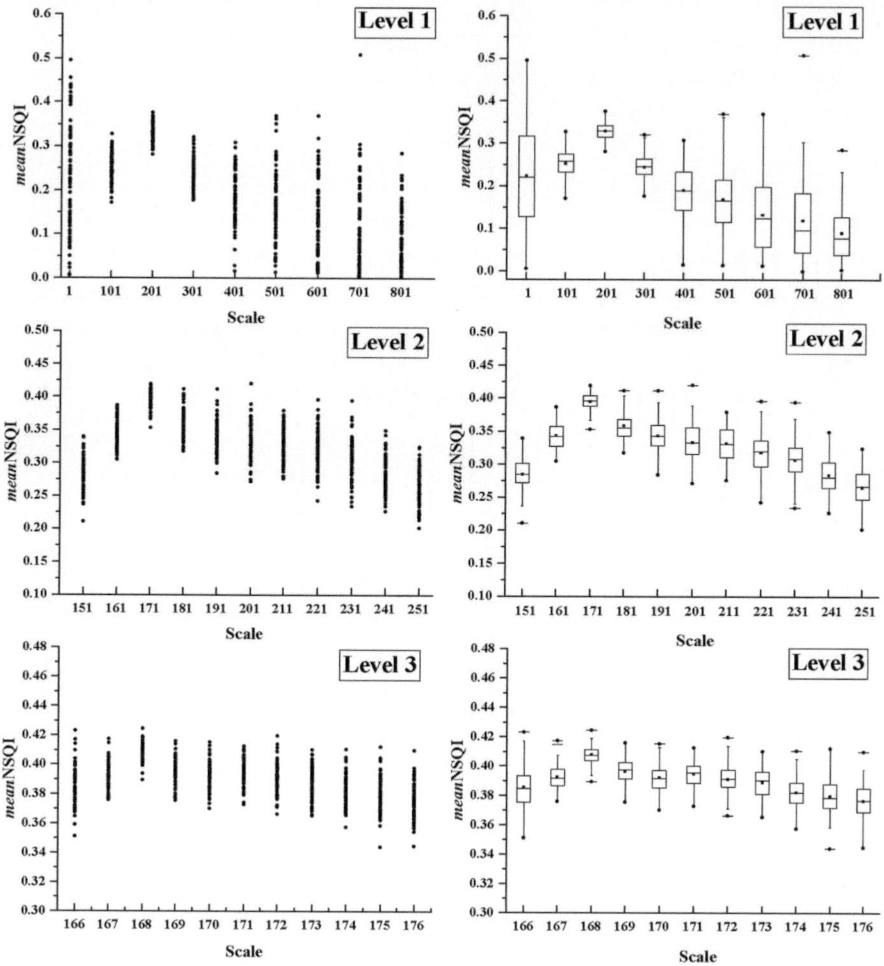


**Fig. 3.7** Distribution and boxplot of meanNSQI at different scales for each level of the WV-2_2013 image

Figure 3.10 describes the optimal scale selection situation for the FWV-3_2014 image, based on three levels and meanNSQI. Similar to the FWV-2_2010 image, the optimal scales for the first two levels are the same, and the final scale chosen for level 3 is scale210, which is very close to the 208 chosen for FWV-2_2010. This indicates that these two images have the potential to be segmented together. At the same time, the changes in meanNSQI at each scale under the three levels of FWV-3_2014 are very frequent and irregular, which is slightly different from the previous images. Looking at the image itself, it can be seen that the FWV-3_2014 image has a higher spatial resolution (0.34m), richer object details, and the size of the segmentation object is more easily affected by the size of the segmentation scale.

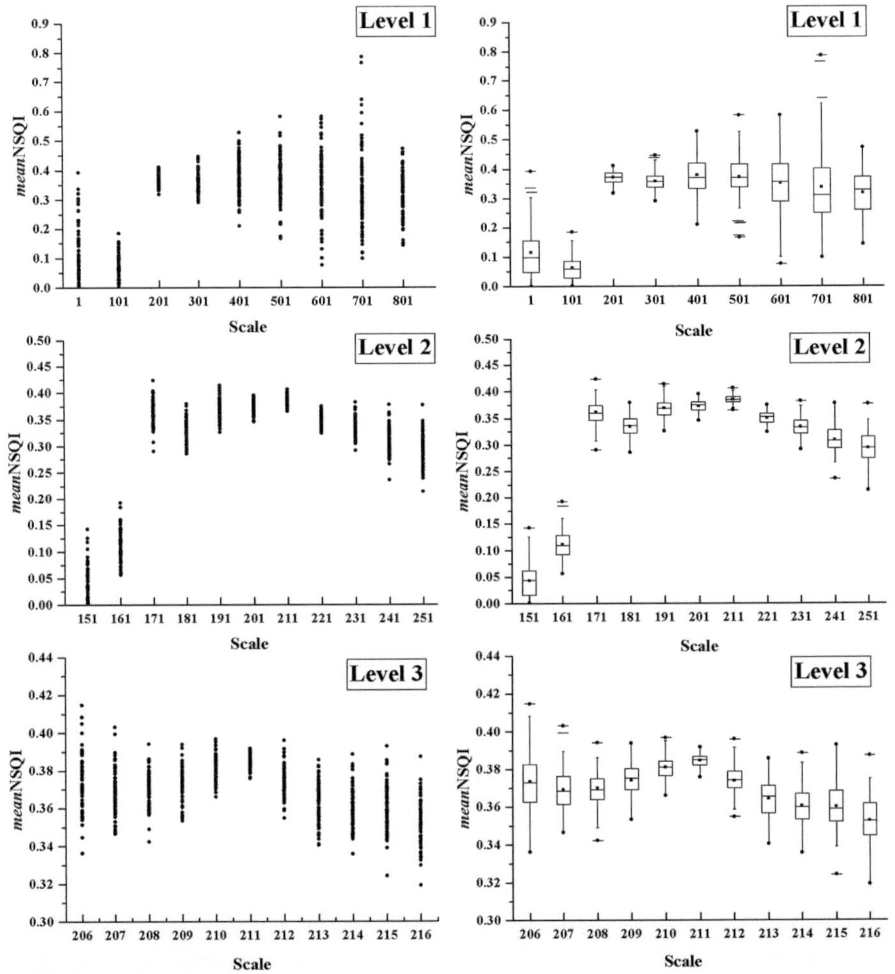


**Fig. 3.8** Distribution and boxplot of meanNSQI at different scales for each level of the FSPOT-6_2015 image

Figure 3.11 describes the optimal segmentation scale selection for the FWV-3_2014 image, based on a two-level and meanNSQI approach. Unlike previous images, the ZY-3_2015 image has a lower spatial resolution (5.8 m), resulting in a significant reduction in the number of pixels for the same area. Therefore, a two-level approach is used for segmentation parameter selection. The scale selection range for level 1 is [1:101], with a step size of 10. The meanNSQI average of scale21 is higher than other scales, so the scale selection range for level 2 is [16:26], with a step size of 1, resulting in an optimal scale of 24. Comparing with the first two images, the segmentation scale of ZY-3_2015 is significantly different. For images of the same area with different resolutions, the optimal scales selected by meanNSQI show significant differences.

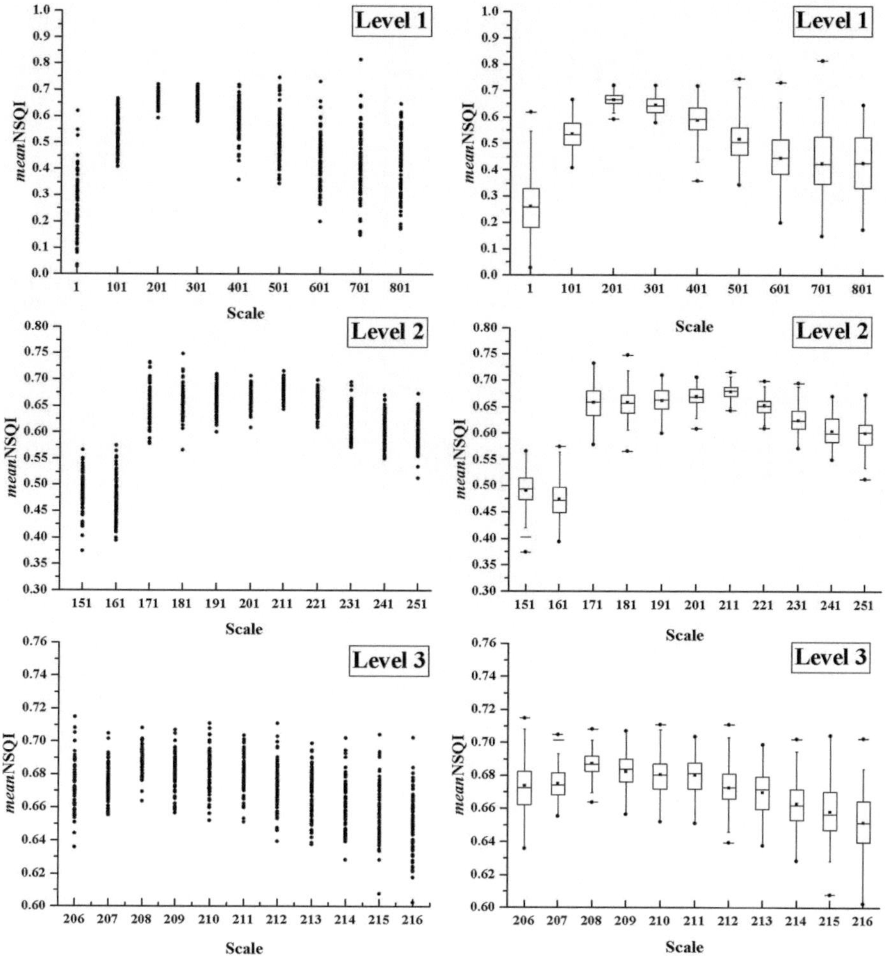


**Fig. 3.9** Distribution and boxplot of meanNSQI at different scales for each level of the FWV-2_2010 image

The spatial resolution of the image will to some extent affect the choice of the optimal segmentation scale, but whether there is a certain correlation or pattern needs further discussion.

Table 3.1 describes the results of the optimal segmentation scale selection for each image in the two experimental areas, as well as the corresponding statistical measures of meanNSQI. For Area 1, the three images have the same spatial resolution. The optimal segmentation scales for WV-2_2012 and WV-2_2013 are basically the same, but the optimal segmentation scale for FSPOT-6_2015 is different. However, in terms of the statistical results of meanNSQI, the segmentation effects of the three images under the optimal segmentation scale are basically the same, and the statistical values of meanNSQI are also quite close. The reasons for this may be the

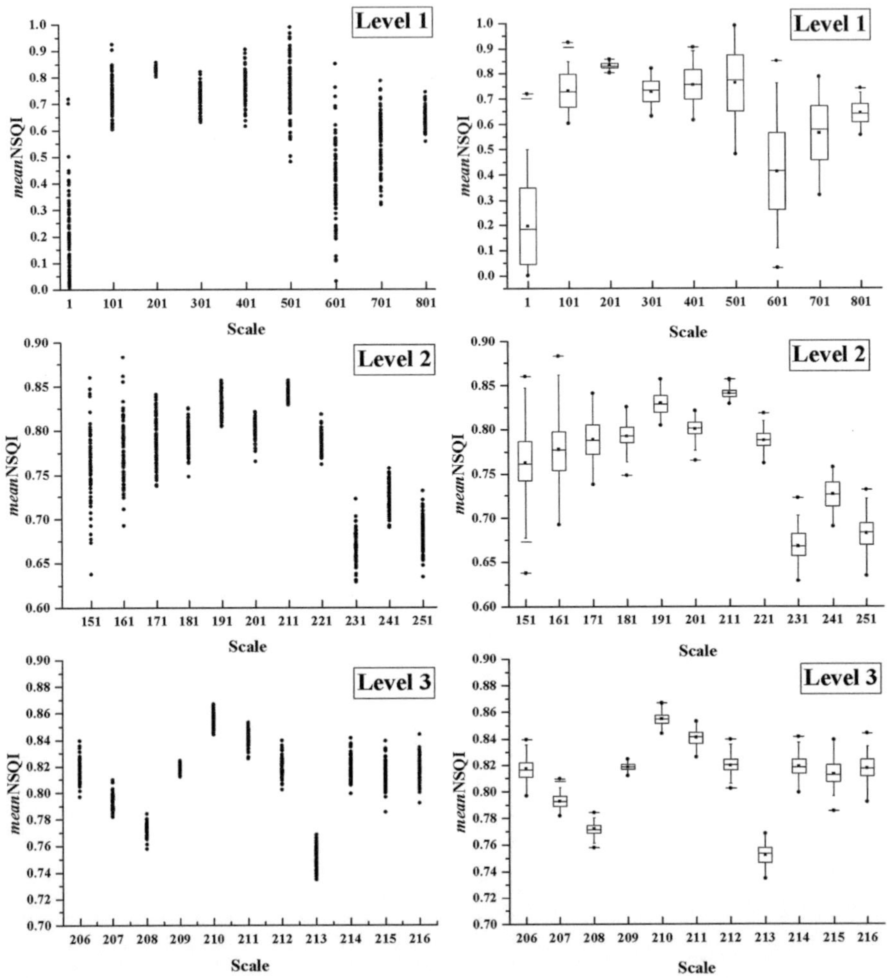


**Fig. 3.10** Distribution and boxplot of meanNSQI at different scales for each level of the FWV-3_2014 image

differences caused by different times, and the differences between panchromatic and multispectral fusion images and ordinary multispectral images. Although the spatial resolution of the fusion image has improved, the detail information of the image itself is still insufficient compared to the non-fused image of the same resolution, so a higher segmentation scale is needed to improve the internal homogeneity of the segmentation object. The relatively consistent statistical results of meanNSQI indicate that the segmentation parameter selection method based on meanNSQI will not be affected by data fusion, proving that although the processing method of improving spatial resolution through fusion will affect the selection of the optimal scale, it will not affect the image segmentation effect under the optimal segmentation scale.

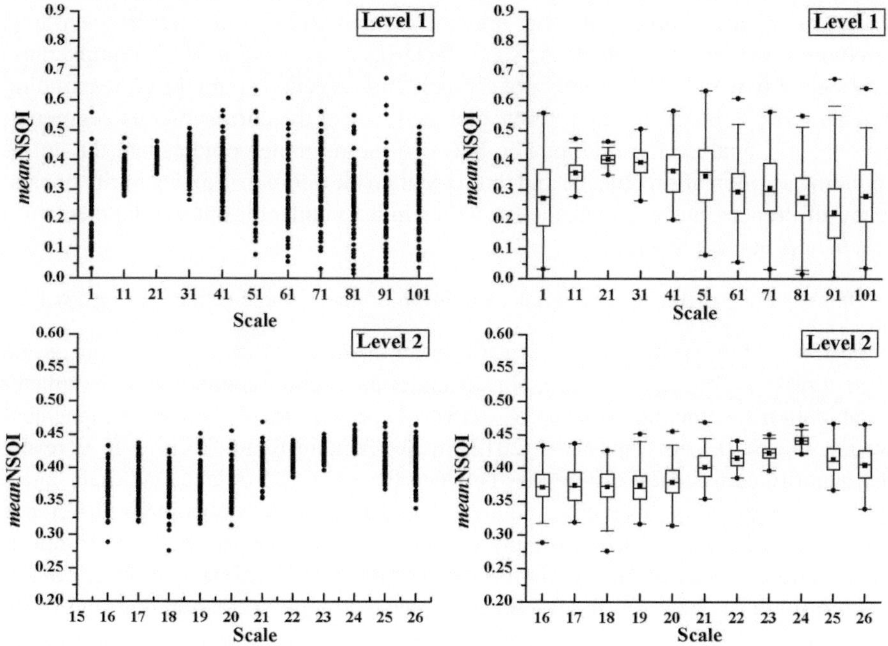


**Fig. 3.11** Distribution and boxplot of meanNSQI at different scales for each level of the ZY-3_2015 image

**Table 3.1** Statistics of segmentation scale selection based on meanNSQI

| Area | Image data | Optimal segmentation scale | meanNSQI | | | |
|---|---|---|---|---|---|---|
| | | | Mean | Standard deviation | Maximum | Minimum |
| Area 1 | WV-2_2012 | 173 | 0.3895 | 0.0064 | 0.4165 | 0.3752 |
| | WV-2_2013 | 168 | 0.4079 | 0.0059 | 0.4241 | 0.3894 |
| | FSPOT-6_2015 | 211 | 0.3846 | 0.0065 | 0.4081 | 0.3758 |
| Area 2 | FWV-2_2010 | 208 | 0.7877 | 0.0088 | 0.8084 | 0.7609 |
| | FWV-3_2014 | 210 | 0.8555 | 0.0042 | 0.8668 | 0.8438 |
| | ZY-3_2015 | 24 | 0.4413 | 0.0081 | 0.4645 | 0.4218 |

### *3.4.3 Segmentation Evaluation Based on Consistency Difference Analysis*

The analysis evaluation includes the following two steps: ① Based on the optimal segmentation scale, select the parameter combination corresponding to the maximum meanNSQI from the 81 parameter combinations at each level as the optimal segmentation parameters, perform multi-scale segmentation on the image, and obtain the segmentation results under the optimal segmentation parameters at each level (including segmentation scale $S$, shape heterogeneity weight $w_{sh}$, compactness heterogeneity weight $w_{cp}$, number of segmentation objects $N$, and the corresponding meanNSQI); ② Based on the reference objects, select the corresponding segmentation objects, analyse the relationship between the reference objects and the corresponding segmentation objects, and thus calculate the geometric and quantity consistency differences, and calculate the evaluation indices under the optimal segmentation parameters at each level.

1. Consistency difference analysis evaluation of each image object in Region 1

In Fig. 3.12, Figure A describes the overlay relationship of all 26 reference objects and their corresponding segmentation objects under the optimal segmentation parameter combination (segmentation scale $S = 183$, shape weight $w_{sh} = 0.6$, compactness weight $w_{cp} = 0.3$) of the WV-2_2012 image. Among them, Figures B–E reveal the quantity consistency differences between the reference objects and their corresponding segmentation objects, Figures B and E show a one-to-many difference relationship, Figure C shows a many-to-one difference relationship, and Figure E shows a one-to-one difference relationship; Figures b–e reveal the geometric consistency differences between the reference objects and their corresponding segmentation objects, all figures show varying degrees of geometric consistency differences, Figures b and d mainly show under-segmentation overlay relationships, and Figures c and e mainly show over-segmentation overlay relationships. For different land cover types, there are varying degrees of geometric or quantity consistency differences. Combined with the analysis in Table 3.2, it can be seen that the main difference in the artificial grassland area is the quantity consistency difference (ED2 is higher), and the consistency difference in the building area is mainly the geometric consistency difference (ED1 is higher). The indicators of the artificial water area are better than other areas, and the segmentation effect is the best; the QR of the building area is higher than other areas, indicating that compared with other areas, the overlap part of the reference objects and segmentation objects in the building cover area is less.

In Fig. 3.13, Figure A describes the overlay relationship of all 26 reference objects and their corresponding segmentation objects under the optimal segmentation parameter combination ($S = 178$, $w_{sh} = 0.6$, $w_{cp} = 0.3$) of the WV-2_2013 image. Figures B and E show a one-to-many number.

The consistency differences are shown in the figures: Figure C shows the many-to-one quantity consistency differences, Figure D shows the one-to-one quantity consistency differences, Figures B and D show the under-segmented geometric consistency differences, and Figures C and E show the over-segmented geometric consistency differences. Combined with Table 3.3, it can be seen that the consistency differences in the artificial grassland areas are mainly quantity consistency differences (with a higher ED2), while the consistency differences in the building areas are mainly geometric consistency differences (with a higher ED1). The OR of the artificial water bodies is lower than that of the artificial grassland and building areas, indicating better segmentation results; the QR of the building areas is higher than other areas, indicating that compared to other areas, the overlap between the reference objects and segmented objects in the building coverage areas is less.

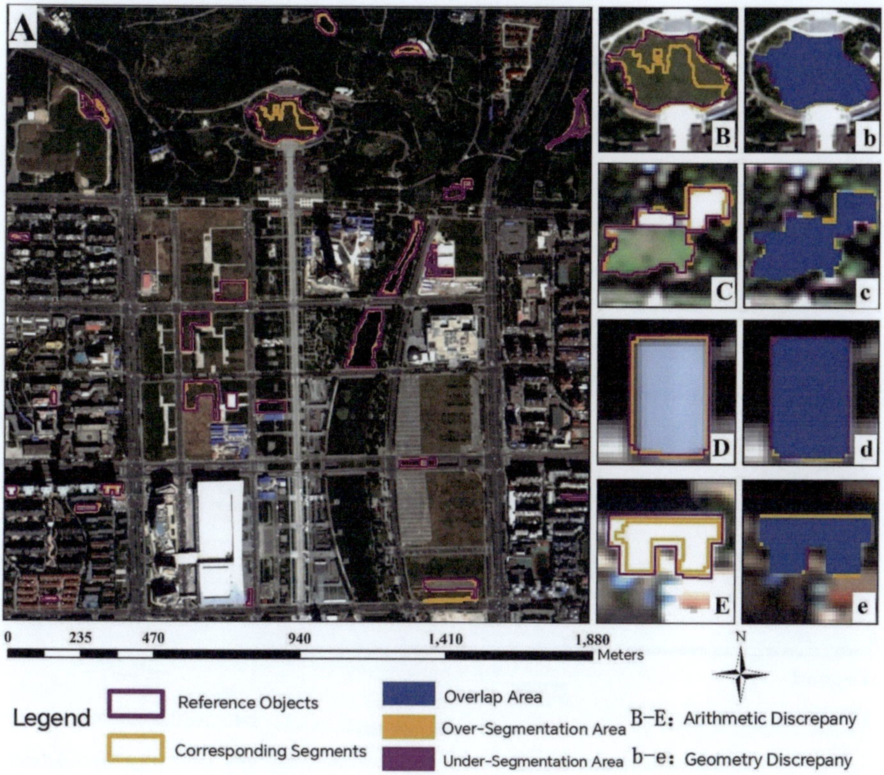


**Fig. 3.12** Correspondence between reference objects and segmentation objects in the WV-2_2012 image (colour image attached at the end)

**Table 3.2** Consistency difference evaluation results of different land cover types in the WV-2_2012 image

| Land cover type | Number of reference objects | Geometric consistency difference evaluation | | | | Quantity consistency difference evaluation | | |
|---|---|---|---|---|---|---|---|---|
| | | QR | OR | UR | ED1 | PSE | NSR | ED2 |
| Artificial grassland | 10 | 0.115 | 0.054 | 0.066 | 0.060 | 0.067 | 0.993 | 0.996 |
| Buildings | 14 | 0.156 | 0.096 | 0.038 | 0.073 | 0.053 | 0.432 | 0.435 |
| Artificial water bodies | 2 | 0.087 | 0.045 | 0.046 | 0.046 | 0.080 | 0.000 | 0.080 |
| Global | 26 | 0.126 | 0.069 | 0.051 | 0.061 | 0.064 | 0.570 | 0.574 |

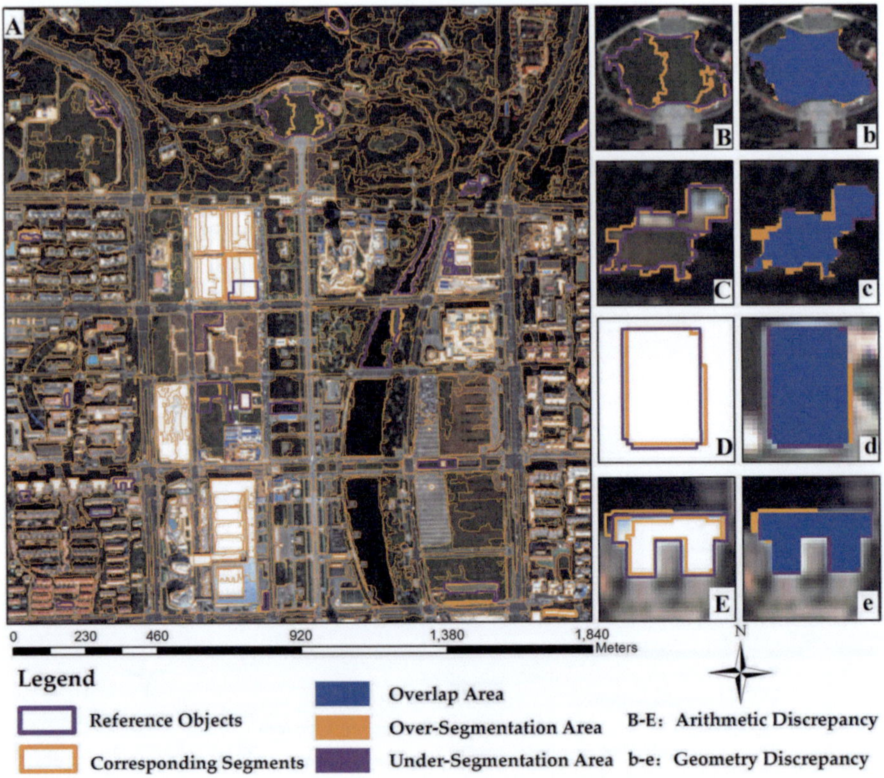


**Fig. 3.13** Correspondence between the reference objects and segmented objects in the WV-2_2013 image (colour image attached)

**Table 3.3** Consistency difference evaluation results for different land cover types in the WV-2_2013 image

| Land cover type | Number of reference objects | Geometric consistency difference evaluation | | | | Quantity consistency difference evaluation | | |
|---|---|---|---|---|---|---|---|---|
| | | QR | OR | UR | ED1 | PSE | NSR | ED2 |
| Artificial grassland | 10 | 0.098 | 0.052 | 0.052 | 0.052 | 0.053 | 1.033 | 1.035 |
| Buildings | 14 | 0.110 | 0.092 | 0.021 | 0.067 | 0.053 | 0.521 | 0.524 |
| Artificial water bodies | 2 | 0.088 | 0.043 | 0.032 | 0.038 | 0.062 | 0.000 | 0.062 |
| Global | 26 | 0.101 | 0.066 | 0.035 | 0.055 | 0.055 | 0.533 | 0.536 |

In Fig. 3.14, Figure A describes the overlay relationship between all 22 reference objects and their corresponding segmented objects under the optimal parameter combination ($S = 221$, $w_{sh} = 0.7$, $w_{cp} = 0.4$) for the FSPOT-6_2015 image. Figures B and D represent one-to-one quantity consistency differences, Figure C represents many-to-one quantity consistency differences, and Figure E represents one-to-one quantity consistency differences. Figure E mainly shows the overlay relationship of over-segmentation, while Figures B to D indicate that both over-segmentation and under-segmentation overlay relationships exist. The consistency differences in the grassland area and the building area are basically the same (their ED1 and ED2 are very close), the indicators of the artificial water area are better than other areas, and the segmentation effect is good; the QR of the building area is higher than other areas, indicating that compared with other areas, the overlap between the reference objects and the segmented objects in the building coverage area is less (Table 3.4).

Considering the three images of Region 1 (Table 3.5), the optimal segmentation scales of the two WV-2 images are relatively close, and the differences in various indicators are also small. However, there is a certain gap with the segmentation parameter optimisation results of FSPOT-6. The spatial resolution of the three images is

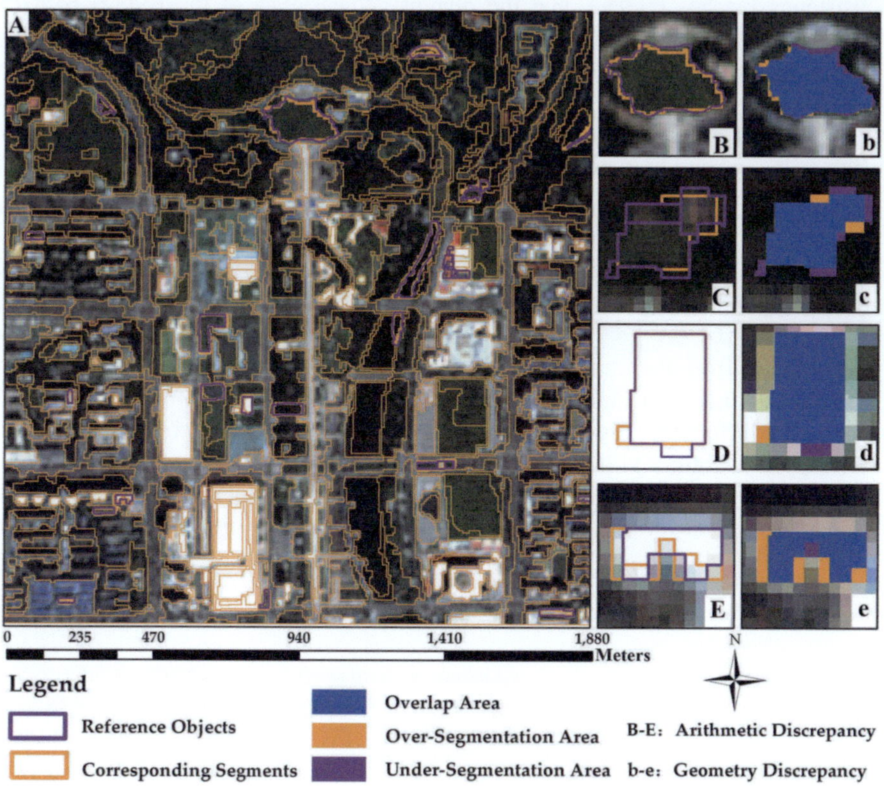


**Fig. 3.14** Correspondence between reference objects and segmented objects in the FSPOT-6_2015 image (colour version in appendix)

**Table 3.4** Consistency difference evaluation results of different land cover types in the FSPOT-6_2015 image

| Land cover type | Number of reference objects | Geometric consistency difference evaluation | | | | Quantity consistency difference evaluation | | |
|---|---|---|---|---|---|---|---|---|
| | | QR | OR | UR | ED1 | PSE | NSR | ED2 |
| Artificial grassland | 8 | 0.152 | 0.041 | 0.127 | 0.094 | 0.044 | 0.519 | 0.520 |
| Buildings | 12 | 0.164 | 0.125 | 0.050 | 0.095 | 0.126 | 0.511 | 0.516 |
| Artificial water bodies | 2 | 0.122 | 0.075 | 0.088 | 0.082 | 0.078 | 0.000 | 0.078 |
| Global | 22 | 0.149 | 0.081 | 0.088 | 0.085 | 0.084 | 0.333 | 0.344 |

the same, but the number of segmentation objects under their respective optimal segmentation scales is quite different. Comparing the consistency difference evaluation results of the three images (Fig. 3.15a), the meanNSQI of the three is basically consistent. The geometric consistency of the WV-2_2013 image objects is better than the other two (the smallest ED1 index is 0.055, QR is 0.101, that is, the geometric consistency reaches 89.9%), and the quantity consistency of the FSPOT-6_2015 image objects is better than the other two (both ED2 and NSR indices are the lowest). Comparing the consistency difference evaluation of different surface cover/land use type objects within the region (Fig. 3.15b), it can be seen that artificial water bodies are superior in all aspects, artificial grassland areas have more quantity consistency differences, and building areas have more geometric consistency differences.

2. Consistency difference analysis and evaluation of each image object in region 2

Figure 3.16, Figure A depicts the overlay relationship of all 31 reference objects and their corresponding segmented objects under the optimal parameter combination ($S = 218$, $w_{sh} = 0.4$, $w_{cp} = 0.5$) for the FWV-2_2010 image. Figures B and C show two examples from the farmland area, while Figures D and E show two examples from the building area. Specifically, Figure B represents a one-to-one numerical consistency difference, Figure C represents a one-to-many numerical consistency difference, and Figures D and E represent many-to-one numerical consistency differences. The geometric consistency difference in Figure B is relatively small, Figures C and D represent over-segmentation geometric consistency differences, and Figure

**Table 3.5** Segmentation parameter optimisation and consistency difference evaluation results of each image in Region 1

| Data | Segmentation parameter optimisation results | | | | | Geometric consistency difference evaluation | | | | Quantity consistency difference evaluation | | |
|---|---|---|---|---|---|---|---|---|---|---|---|---|
| | $S$ | $w_{sh}$ | $w_{cp}$ | $N$ | mNSQI | QR | OR | UR | ED1 | PSE | NSR | ED2 |
| D1 | 173 | 0.6 | 0.4 | 792 | 0.417 | 0.126 | 0.069 | 0.051 | 0.061 | 0.064 | 0.570 | 0.574 |
| D2 | 168 | 0.6 | 0.3 | 810 | 0.424 | 0.101 | 0.066 | 0.035 | 0.055 | 0.055 | 0.533 | 0.536 |
| D3 | 211 | 0.7 | 0.4 | 554 | 0.408 | 0.149 | 0.081 | 0.088 | 0.085 | 0.084 | 0.333 | 0.344 |

*Note* D1 is WV-2_2012, D2 is WV-2_2013, D3 is FSPOT-6_2015, mNSQI refers to meanNSQI

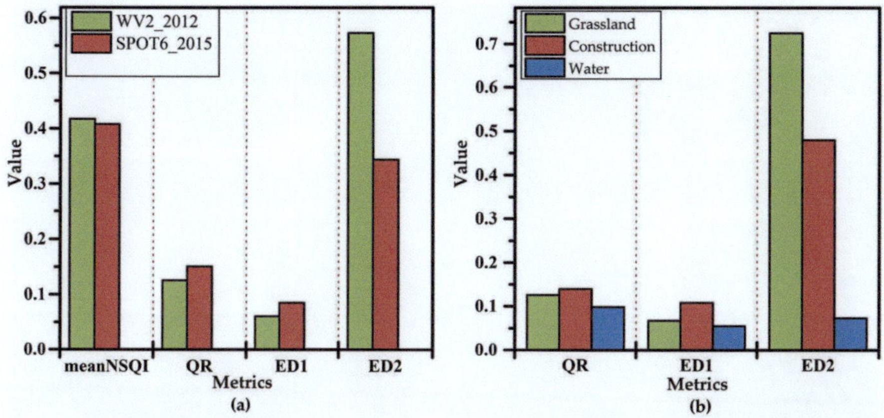


**Fig. 3.15** Comparison of consistency difference evaluation results of each image object and each surface cover type in Region 1

E represents under-segmentation geometric consistency differences. Combined with Table 3.6, the analysis of consistency evaluation for different land use/land cover type areas reveals that the consistency differences in the three areas are similar, with all indicators being relatively close. The main consistency difference in the farmland area is the numerical consistency difference (higher ED2), while the main consistency difference in the building area is the geometric consistency difference (higher ED1). The OR of the three areas is also similar, with a large overlap between the reference objects and segmented objects in each area, resulting in similar segmentation effects. Figure 3.17, Figure A depicts the overlay relationship of all 30 reference objects and their corresponding segmented objects under the optimal parameter combination (S = 220, $w_{sh} = 0.4$, $w_{cp} = 0.3$) for the FWV-3_2014 image. Figures B and C show the farmland area…

Two cases are presented for each domain, with Figures D and E showing two cases in the building area. Specifically, Figure B represents a one-to-one numerical consistency difference, Figure C represents a one-to-many numerical consistency difference, and Figures D and E represent many-to-one numerical consistency differences. The geometric consistency difference in Figure B is relatively small, Figure C represents an under-segmented geometric consistency difference, and Figures D and E represent over-segmented geometric consistency differences. In conjunction with Table 3.7, the consistency difference in the farmland area is primarily a numerical consistency difference (higher ED2), while the consistency difference in the building area is primarily a geometric consistency difference (higher ED1). The OR of the river is similar to that of the farmland area, indicating good segmentation results; the QR of the residential building area is higher than other areas, suggesting that compared to other areas, the overlap between the reference object and the segmented object in the building coverage area is less.

In Fig. 3.18, Figure A describes the ZY-3_2015 image under the optimal segmentation parameter combination ($S = 24$, $w_{sh} = 0.2$, $w_{cp} = 0.3$) The overlay relationships of all 24 reference objects and their corresponding segmented objects are shown. Figures B and C illustrate two cases in the farmland area, while Figures

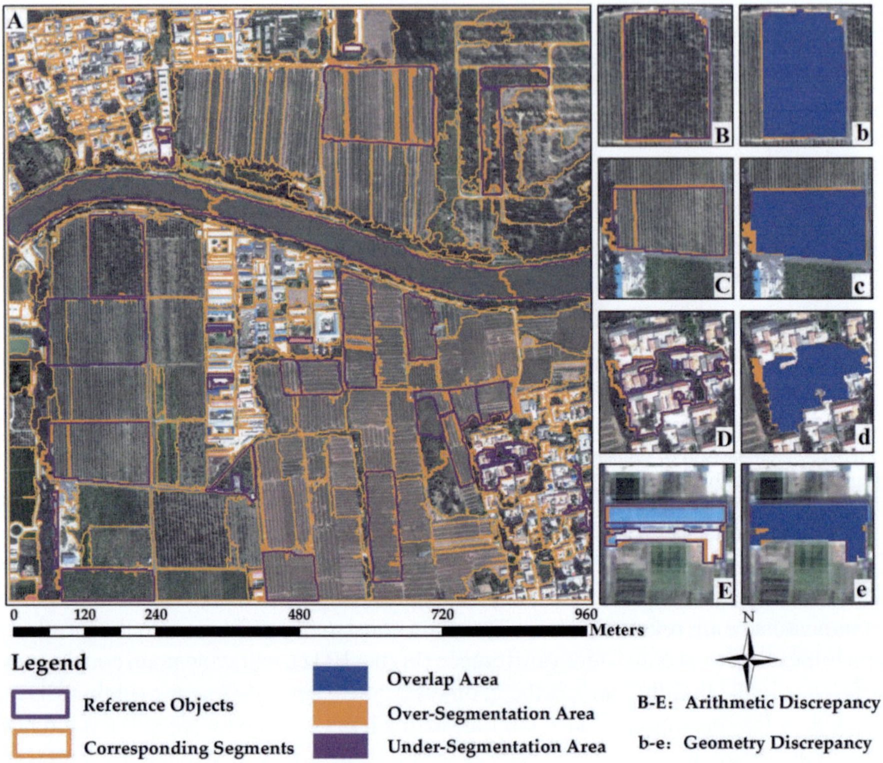


**Fig. 3.16** FWV-2_2010 image reference objects and segmented objects corresponding relationship (colour image attached later)

**Table 3.6** Consistency difference evaluation results for different area types of FWV-2_2010 image

| Area type | Number of reference objects | Geometric consistency difference evaluation | | | | Numerical consistency difference evaluation | | |
|---|---|---|---|---|---|---|---|---|
| | | QR | OR | UR | ED1 | PSE | NSR | ED2 |
| Farmland | 15 | 0.096 | 0.018 | 0.050 | 0.038 | 0.047 | 0.679 | 0.681 |
| Residential area | 14 | 0.095 | 0.055 | 0.041 | 0.048 | 0.116 | 0.567 | 0.579 |
| River | 2 | 0.085 | 0.021 | 0.047 | 0.036 | 0.079 | 0.652 | 0.657 |
| Global | 31 | 0.091 | 0.034 | 0.046 | 0.042 | 0.081 | 0.629 | 0.634 |

D and E depict two cases in the built-up area. Specifically, Figure B represents a one-to-one numerical consistency difference, Figure C a one-to-many numerical consistency difference, and Figures D and E a many-to-one numerical consistency difference. Figure B exhibits good geometric consistency, Figure C primarily shows under-segmentation overlay relationships, and Figures D and E mainly display over-segmentation overlay relationships. An analysis of the consistency evaluation for

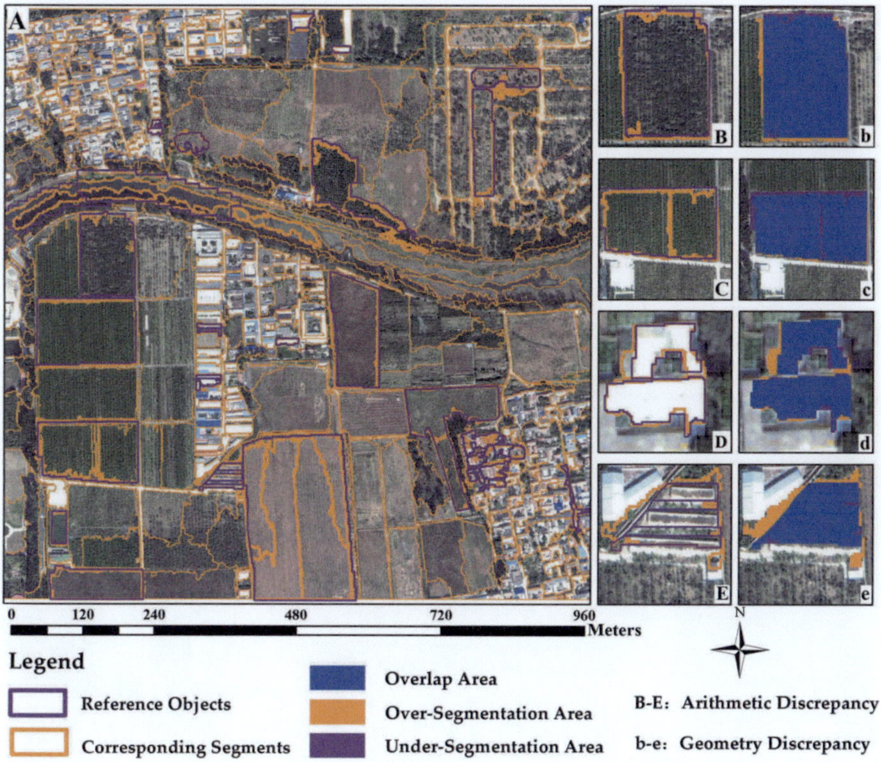


**Fig. 3.17** Relationship between the reference object and the segmented object in the FWV-3_2014 image (colour image attached)

**Table 3.7** Presents the consistency difference evaluation results for different land cover types in the FWV-3_2014 image

| Land cover type | Number of reference objects | Geometric consistency difference evaluation | | | | Numerical consistency difference evaluation | | |
|---|---|---|---|---|---|---|---|---|
| | | QR | OR | UR | ED1 | PSE | NSR | ED2 |
| Farmland | 13 | 0.077 | 0.044 | 0.026 | 0.036 | 0.075 | 1.000 | 1.003 |
| Buildings | 14 | 0.148 | 0.063 | 0.086 | 0.075 | 0.047 | 0.333 | 0.337 |
| River | 3 | 0.075 | 0.011 | 0.047 | 0.034 | 0.059 | 0.372 | 0.377 |
| Global | 30 | 0.105 | 0.036 | 0.063 | 0.051 | 0.060 | 0.608 | 0.611 |

different land cover types reveals that the main consistency difference in agricultural areas is numerical consistency difference (higher ED2), while in built-up areas, it is geometric consistency difference (higher ED1). The metrics for rivers outperform those for farmland and built-up areas, indicating better segmentation results. The QR for the built-up area is higher than for other areas, suggesting that the overlap between the reference objects and segmented objects in the built-up area is less compared to other areas (Table 3.8).

Upon comparing the three images of Area 2 (Table 3.9), the optimal segmentation scales of FWV-2 and FWV-3 are relatively close, with minor differences in various indicators, while the segmentation parameter optimisation results of ZY-3_2015 show a significant disparity. As shown in Fig. 3.19a, the evaluation indicators of

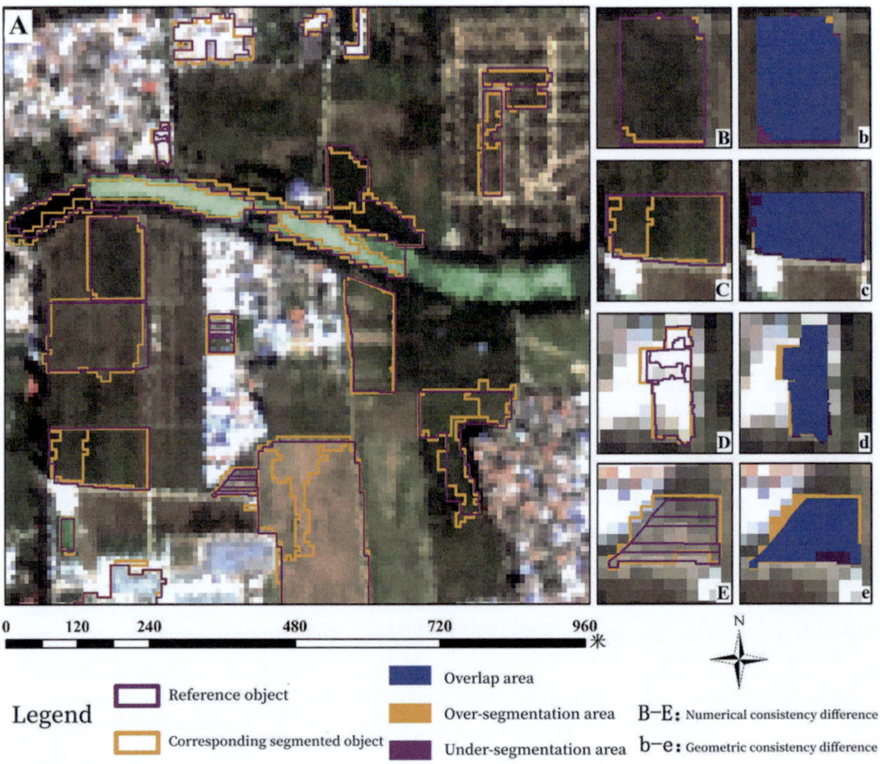


**Fig. 3.18** Correspondence between reference objects and segmented objects in ZY-3_2015 imagery (colour image attached)

**Table 3.8** Consistency difference evaluation results for different land cover types in ZY-3_2015 imagery

| Area type | Number of reference objects | Geometric consistency difference evaluation | | | | Numerical consistency difference evaluation | | |
|---|---|---|---|---|---|---|---|---|
| | | QR | OR | UR | ED1 | PSE | NSR | ED2 |
| Farmland | 10 | 0.125 | 0.155 | 0.08 | 0.123 | 0.228 | 0.792 | 0.824 |
| Built-up Area | 11 | 0.213 | 0.16 | 0.133 | 0.147 | 0.08 | 0.5 | 0.506 |
| River | 3 | 0.075 | 0.011 | 0.047 | 0.034 | 0.059 | 0.372 | 0.377 |
| Global | 24 | 0.150 | 0.139 | 0.115 | 0.121 | 0.154 | 0.602 | 0.621 |

FWV-2 and FWV-3 are basically consistent. Due to the lower spatial resolution, the geometric consistency between the segmented objects and reference objects in the ZY-3_2015 image is poor, with fewer overlapping areas. The geometric consistency of the FWV-2_2010 image objects is slightly better than the other two (the smallest ED1 is 0.042, QR is only 0.091, with a geometric consistency of up to 90.9%), and the quantity consistency of the FWV-3_2014 image objects is slightly better than the other two (the smallest ED2 is 0.611, the NSR data of the three are very close). Comparing the evaluation results of different land cover, rivers show less variation in all aspects, farmland shows more quantity consistency differences, and residential buildings show more geometric consistency differences. From the evaluation results of ZY-3_2015, it can be seen that the meanNSQI value of the lower resolution is significantly lower than that of the high resolution, and there are more geometric consistency differences, but the quantity consistency differences are basically the same.

**Table 3.9** Segmentation parameter optimisation and consistency difference evaluation results of each image in area 2

| Data | Segmentation parameter optimisation results | | | | | Geometric consistency difference evaluation | | | | Quantity consistency difference evaluation | | |
|---|---|---|---|---|---|---|---|---|---|---|---|---|
| | *S* | $w_{sh}$ | $w_{cp}$ | *N* | mNSQI | QR | OR | UR | ED1 | PSE | NSR | ED2 |
| D4 | 218 | 0.4 | 0.5 | 419 | 0.808 | 0.091 | 0.034 | 0.046 | 0.042 | 0.081 | 0.629 | 0.634 |
| D5 | 220 | 0.4 | 0.3 | 401 | 0.867 | 0.105 | 0.036 | 0.063 | 0.051 | 0.060 | 0.608 | 0.611 |
| D6 | 221 | 0.2 | 0.3 | 165 | 0.465 | 0.150 | 0.119 | 0.135 | 0.121 | 0.154 | 0.602 | 0.621 |

*Note* D4 is FWV-2_2010; D5 is FWV-3_2014; D6 is ZY-3_2015; mNSQI is meanNSQI

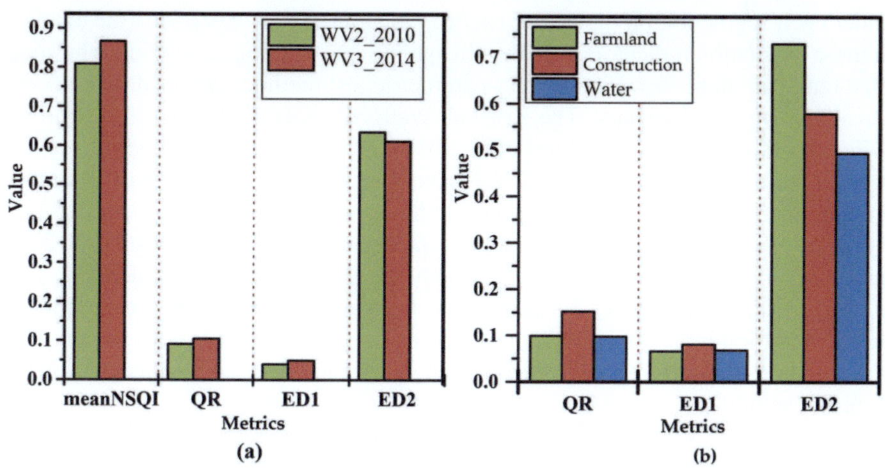


**Fig. 3.19** Consistency difference evaluation results of each image object and each land cover type in Area 2

## 3.5 Discussion

The experimental results show that multi-scale segmentation based on meanNSQI optimised segmentation parameters can yield segmented objects that maintain a high degree of geometric and numerical consistency with the reference objects. However, some issues were also identified, such as the sensitivity of consistency difference evaluation indicators, which warrants further discussion. When the segmentation scale changes by a certain magnitude, the changes in the related consistency evaluation indicators (mainly ED1 and ED2) are analysed to assess the reliability of ED1 and ED2 as evaluation indicators.

Furthermore, the multi-scale segmentation in this chapter is intended to provide reasonable image objects for subsequent object-based change detection (OBCD). Therefore, the consistency issues brought about by the overlay of multi-temporal images in OBCD also need to be considered. That is, when two remote sensing images covering the same area and with the same resolution are segmented at the same scale using band overlay, are the optimised segmentation parameters consistent with the previous ones? Is there a significant consistency difference between the segmented objects of the overlaid images and those obtained from a single image?

### *3.5.1 Sensitivity Analysis of Consistency Difference Evaluation Indicators*

Firstly, based on the previously obtained optimal segmentation parameters, an appropriate range is selected to obtain the distribution of the main evaluation indicators under each segmentation parameter (mainly ED1 and ED2). Secondly, the changes in these evaluation indicators with the change in segmentation scale within the specified range are observed. Finally, non-parametric test methods are used to verify the sensitivity of each indicator. The Kruskal–Wallis test (Kruskal et al. 1952) and the Wilcoxon rank-sum test (Gibbons et al. 2010) are chosen to verify the differences in each indicator across different scale pairs. The Kruskal–Wallis test can examine whether there are significant differences in the distributions of multiple populations, while the Wilcoxon rank-sum test judges whether there are differences in the distributions of the two populations from which the paired samples are derived. That is, the Kruskal–Wallis test verifies whether there are differences in the samples of these evaluation indicators within the specified range, while the Wilcoxon rank-sum test compares the differences in each paired sample to determine the location of the differing samples.

Area one's WV-2_2013 is selected as the experimental object. As shown in Fig. 3.7, the optimal segmentation parameter group is scale178, so the sample range is set to [68:288], with a step size of 10. The mean values of ED1 and ED2 for each parameter group are calculated, and their change curves with the segmentation scale are shown in Fig. 3.20.

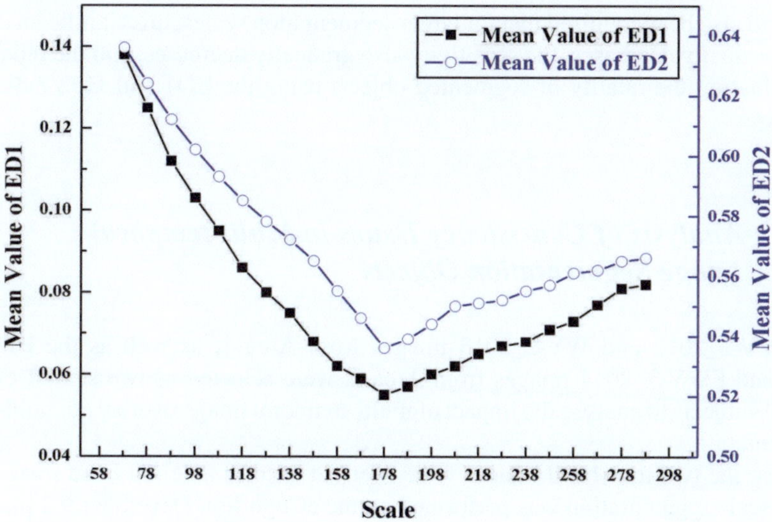


**Fig. 3.20** Change curves of mean ED1 and ED2 with segmentation scale

As shown in Fig. 3.20, the mean values of ED1 and ED2 change with the segmentation scale, reaching a minimum at scale178. Before scale178, there is a significant continuous decrease; after scale178, there is a slight increase. This shows that the optimal segmentation scale of scale178 selected based on meanNSQI is consistent with this, and the change in evaluation indicators for segmentation parameter combinations less than scale178 is more obvious and easier to distinguish. However, the change in evaluation indicators obtained from segmentation parameter combinations greater than scale178 is slightly slower. Therefore, it is necessary to conduct a significance test on it to analyse the sensitivity of the evaluation indicators to changes in segmentation scale parameters. Firstly, the Kruskal–Wallis test is used to compare the differences in all scale parameter groups as a whole, and then the Wilcoxon rank-sum test is used to compare and analyse the differences in each scale paired sample. The significance level for both hypothesis tests is set at 0.05. Paired samples are selected according to a step size of 10, i.e., scale68–78, scale78–88, …, scale278–288.

As shown in Table 3.10, the Kruskal–Wallis test results for ED1 and ED2 show significant differences in the overall samples at a significance level of 0.05; the Wilcoxon rank-sum test results show that there are significant differences in each paired sample when the segmentation scale is less than 198, and the ED1 and ED2 indicators can be used to distinguish each group of samples. However, when the segmentation scale is greater than 198, the differences in each paired sample are no longer significant. The previously optimised segmentation scale was scale178. From the above conclusions, it can be seen that the ED1 and ED2 indicators have a certain correlation with the quality of the segmented objects under different segmentation scale parameters, and they are quite sensitive to changes in segmentation scale. There are significant differences in the sample data, and the quality of the segmented objects evaluated by the ED1 and ED2 indicators is reliable and effective. However, it should also be recognised that for larger segmentation scales, this significant difference gradually decreases, the sensitivity also gradually decreases, and the reliability of evaluating the quality of segmented objects using the ED1 and ED2 indicators decreases.

### *3.5.2 Analysis of Consistency Issues in Multi-temporal Image Segmentation Objects*

The WV-2_2012 and WV-2_2013 images from Area 1, as well as the FWV-2_2010 and FWV-3_2014 images from Area 2, were selected as two sets of experimental subjects to analyse the impact of multi-temporal image overlay on multi-scale segmentation.

First, the two four-band images were overlaid to form an eight-band image, and multi-scale segmentation was performed on the eCognition Developer 9.2 platform (all band weights were set to 1). The optimal segmentation parameters were selected using the three-layer level and meanNSQI method, and the results of the two sets of experiments are shown in Table 3.11.

As can be seen from Table 3.11, the segmentation parameters selected from the overlay images are similar to the original two images. However, due to the spectral or shape changes in the multi-temporal image objects, the segmentation objects of the overlay images are more complex and detailed compared to the original two image objects. Therefore, the optimal segmentation scale is slightly smaller than the other two images, with more segmentation objects than the original images, and a smaller meanNSQI.

In order to analyse the overlay differences between the overlay image and the original image segmentation objects, the above consistency difference analysis method is used for comparison. Using the segmentation objects of the original image as reference objects, and the segmentation objects of the overlay image as corresponding segmentation objects, the geometric and quantity consistency differences between the overlay image and the original image segmentation objects are obtained by calculating various evaluation indices.

Figure 3.21 shows that there is not much difference in the geometric and quantity consistency difference evaluation between the image objects of each region and the overlay image, and the quantity consistency difference is more obvious (ED2 is higher).

From the above analyses, it can be seen that whether it is Region One or Region Two, in the multi-scale segmentation process of OBCD, the overlay image has little effect on the selection of segmentation parameters, but the number of segmentation objects and meanNSQI are slightly different from the original image. In the comparison analysis of consistency differences between overlay image and original segmentation objects, the evaluation indicators reveal that the overlay image and the original image segmentation objects have relatively consistent boundaries and high geometric consistency, but there are differences in quantity consistency.

**Table 3.10** Results of sensitivity analysis of ED1 and ED2 to segmentation scale

| Indicator | *P*-value of Kruskal–Wallis test | *P*-value of Wilcoxon rank-sum test *P* | | | | | | | | | |
|---|---|---|---|---|---|---|---|---|---|---|---|
| | | 68–78 | 78–88 | 88–98 | 98–108 | 108–118 | 118–128 | 128–138 | 138–148 | 148–158 | 158–168 |
| ED1 | < 0.0001 | < 0.0001 | < 0.0001 | < 0.0001 | 0.0029 | 0.0068 | 0.0097 | 0.0306 | 0.0899 | 0.1162 | 0.2289 |
| ED2 | < 0.0001 | < 0.0001 | < 0.0001 | < 0.0001 | 0.0045 | 0.0088 | 0.0157 | 0.0428 | 0.0962 | 0.184 | 0.281 |

| Indicator | Wilcoxon rank-sum test *P* values | | | | | | | | | | | |
|---|---|---|---|---|---|---|---|---|---|---|---|---|
| | 168–178 | 178–188 | 188–198 | 198–208 | 208–218 | 218–228 | 228–238 | 238–248 | 248–258 | 258–268 | 268–278 | 278–288 |
| ED1 | 0.346 | 0.692 | 0.904 | 1 | 1 | 1 | 1 | 0.9456 | 1 | 1 | 1 | 0.9065 |
| ED2 | 0.451 | 0.730 | 0.981 | 1 | 1 | 1 | 0.9874 | 1 | 0.8879 | 1 | 0.9564 | 1 |

*Note* The significance level $\alpha = 0.05$

**Table 3.11** Based on ***mean*** NSQI segmentation scale selection statistical results

| Region | Image data | Optimal segmentation parameters | | | Number of segmentation objects | meanNSQI |
|---|---|---|---|---|---|---|
| | | $S$ | $w_{sh}$ | $w_{cp}$ | | |
| Region one | WV-2_2012 | 183 | 0.6 | 0.4 | 792 | 0.4165 |
| | WV-2_2013 | 178 | 0.6 | 0.4 | 810 | 0.4241 |
| | Overlay1 | 166 | 0.6 | 0.3 | 826 | 0.3948 |
| Region two | FWV-2_2010 | 218 | 0.4 | 0.5 | 419 | 0.8084 |
| | FWV-3_2014 | 220 | 0.4 | 0.3 | 401 | 0.8668 |
| | Overlay2 | 196 | 0.4 | 0.4 | 432 | 0.7845 |

*Note* Overlay1 refers to the overlay image of WV-2_2012 and WV_2013; Overlay2 refers to the overlay image of FWV-2_2010 and FWV-3_2014

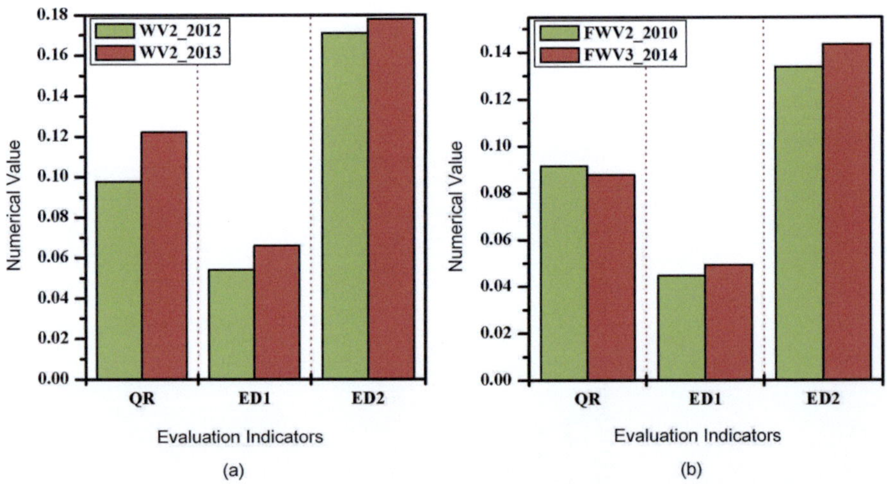


**Fig. 3.21** Analysis results of consistency differences between overlay image and original image segmentation objects

## 3.6 Summary

This chapter, based on the multi-scale segmentation algorithm, proposes a segmentation quality evaluation index (NSQI) that comprehensively considers the homogeneity within the image object and the heterogeneity outside. It uses a hierarchical screening method based on meanNSQI to optimise the segmentation parameters, and evaluates the optimisation of segmentation parameters based on the geometric and numerical consistency differences between the reference object and the corresponding segmented object. The research content and conclusions are as follows.

(1) The construction process of NSQI, the optimisation process of segmentation parameters based on meanNSQI and hierarchical screening are proposed. Under the principle of maximising the internal homogeneity and external heterogeneity of the image object, the NSQI index is constructed, and the multi-band meanNSQI index is obtained; a method of MRS segmentation parameter optimisation using hierarchical screening combined with the meanNSQI index is designed. The results of six experiments in two regions show that the meanNSQI of the three images in region one is very close (0.417, 0.424 and 0.408 respectively), with WV-2_2013 slightly higher; in region two, the meanNSQI of ZY-3_2015 in the three images shows a significant difference (0.808, 0.867 and 0.465 respectively). Combining these images' spatial resolution, it is known that this method results in different segmentation parameter selections for remote sensing images of different resolutions, and the image enhancement method through panchromatic fusion will affect the selection of segmentation parameters to some extent.

(2) A method for evaluating the optimisation of segmentation parameters based on consistency difference analysis is proposed. Based on the analysis of the overlap relationship between the reference object and the corresponding segmented object, the geometric and numerical consistency of the reference object and the corresponding segmented object are evaluated using evaluation indicators such as ED1 and ED2, measuring the quality of the segmented object based on the MRS algorithm, and thus evaluating the optimisation results of its segmentation parameters. The experimental results show that the segmentation results under the optimal segmentation parameters of the six images in the two regions have achieved a lower geometric consistency difference and numerical consistency difference. The geometric consistency of the segmented objects of the six images all reached over 85% (as calculated by QR), ED1 were 0.061, 0.055, 0.085, 0.042, 0.051 and 0.121 respectively, and ED2 were 0.574, 0.536, 0.344, 0.634, 0.611 and 0.621 respectively. It can be seen that this consistency difference analysis can more objectively evaluate the quality of the segmented object, and the overlap relationship between the reference object and the corresponding segmented object can provide a more intuitive evaluation of the segmentation effect of MRS.

(3) The results of the sensitivity analysis of the main evaluation indicators of consistency difference (ED1 and ED2) and segmentation scale show that within a certain segmentation scale range, ED1 and ED2 maintain a stable correlation with the segmentation scale and are sensitive to changes in the segmentation scale. However, if the segmentation scale is too large, this sensitivity will weaken, but generally such large-scale segmentation parameters will not be used as the optimal segmentation parameters. Therefore, it can be proven that the optimisation evaluation of segmentation parameters based on consistency difference analysis is reliable.

(4) The use of image overlay in OBCD can ensure that image objects at different times have the same boundaries and numbers, and this method of image overlay will not have a significant impact on the optimisation results of segmentation

parameters. However, compared with the original single image segmentation object, the overlaid image object has a small geometric consistency difference and a larger numerical consistency difference, indicating that the overlaid image segmentation object can maintain a relatively consistent boundary with the original single image segmentation object, but there is a difference in the number of objects.

# Chapter 4
# Image Object Multi-feature Selection

Based on the generated image objects, this chapter addresses the multi-feature selection problem in the process of image object change detection, and conducts research on the optimal selection of multiple image object features based on the genetic particle swarm optimisation algorithm. On the basis of discussing feature selection theory and the genetic particle swarm optimisation algorithm, and combining the fitness function based on the mean and variance ratio, a multi-feature optimisation step based on the genetic particle swarm is designed. It is further applied to the change detection of image objects, and the convergence, effectiveness, and efficiency of the algorithm are analysed in detail by comparing with other optimisation methods.

In the field of object-oriented change analysis, high-resolution remote sensing images are often segmented into image objects for processing, which enriches the feature space of the image. However, influenced by traditional pixel-level change detection, the spectral features of objects are habitually used as the only input feature for change detection. The significant difference between objects and pixels lies in the richer features of objects. When using object-oriented image analysis ideas, it should be considered that not only spectral feature analysis, but also the fusion analysis of multiple features such as shape features and texture features. However, some analyses believe that the selection of the number of features will have a certain impact on the detection accuracy of the model method (Tokarczyk et al. 2015), so it is the main manifestation of object-level change detection being superior to pixel-level change detection to choose features reasonably for change detection.

In view of the research on high-resolution remote sensing image object change detection in the past, feature selection is mainly carried out through expert knowledge or one-by-one experiments, which is inefficient and has limitations. This book uses the particle swarm algorithm, which is robust and has a simple algorithm theory in swarm intelligence algorithms, and introduces the mutation operation of the genetic algorithm to construct the optimisation algorithm for image object feature selection. The main content of the research is to use the genetic particle swarm algorithm as the search algorithm for feature selection, and to compare and analyse with other search



Q. Chen et al., *Multiple Feature Object-Based Change Detection Technology for Very High-Resolution Remote Sensing Image*, https://doi.org/10.1007/978-981-95-3917-8_4

methods; to choose the Object Based-Hybrid Multivariate Alteration Detection, OB-HMAD, as a method for image object change detection, analyses the accuracy of change detection in various algorithms, and compares the results of multi-feature change detection with single-feature change detection. In addition, it discusses the initial parameters of the algorithm such as the number of candidate features, the initial population size, and their impact on the reliability, convergence, and efficiency of the algorithm.

This chapter is composed of five sections. Section 4.1 covers the theoretical basis of feature selection; Sect. 4.2 discusses feature optimisation selection based on GPSO; Sect. 4.3 presents experiments and result analysis; Sect. 4.4 is a discussion; and Sect. 4.5 concludes the chapter. The research content framework is shown in Fig. 4.1.

## 4.1 Theoretical Basis of Feature Selection

In the field of machine learning, feature selection (FS) is the process of selecting a closely related appropriate subset from the existing features of an object. This subset can be effectively used for the final learning purpose, reducing the time and space complexity of machine learning (Li 2010). The task of feature selection in image object change detection is to select a set of highly discriminative feature subsets (composed of n image object features, $n < N$) from the N-dimensional image object feature space. Essentially, feature selection can be seen as a re-optimisation problem of discrete combinations. Since reasonable feature selection can maximise the preservation of change information in image object features, it is beneficial for the identification and extraction of object change detection. Therefore, feature selection is crucial for improving the efficiency of change detection.

Feature selection algorithms have two main parts: a search algorithm and an evaluation criterion. As analysed in Sect. 1.2.2, some evolutionary search algorithms have gradually become mainstream in feature selection methods, among which swarm intelligence optimisation algorithms have stronger global search capabilities. As a typical representative of swarm intelligence optimisation algorithms, the PSO algorithm is a versatile algorithm suitable for parallel processing.

To further enhance the global search capability and solution speed of PSO, this book introduces the crossover operation from genetic algorithms into the PSO algorithm, merging them into a genetic particle swarm algorithm (GPSO) to solve the problem of optimising feature selection (Chuang et al. 2008; Ghamisi et al. 2015; Inbarani et al. 2014; Moradi et al. 2016; Nazir et al. 2014; Yang et al. 2007). GPSO has the advantages of faster solution speed, simple parameter setting, and stronger global search, but it also has some drawbacks, such as strong randomness leading to irreproducible experiments, strong parameter sensitivity, and the convergence result of the algorithm is easily affected by the initial parameter setting (Moradi et al. 2016). In recent years, research on the application of PSO or hybrid PSO algorithms in the field of feature selection has gradually unfolded (Chuang et al. 2011; Liu

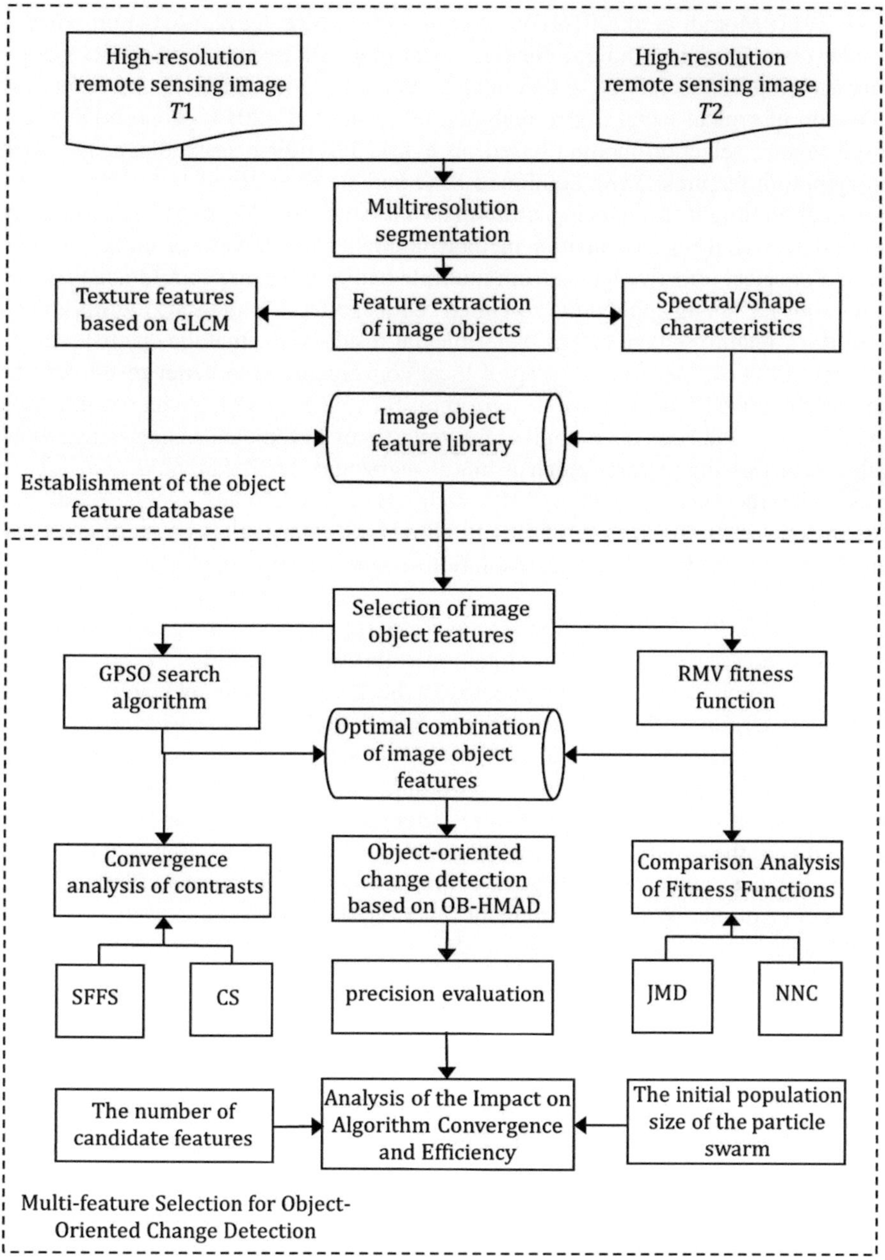


**Fig. 4.1** Research content framework for multi-feature selection of image objects based on genetic particle swarm optimisation algorithm

et al. 2011; Moradi et al. 2016; Wang et al. 2007). Nazir et al. (2014) proposed a gender classification technique for real facial images based on PSO and GA algorithms, using this technique to distinguish gender by extracting and optimising the selection of typical facial image features; Inbarani et al. (2014) proposed a supervised feature selection method based on hybrid PSO for disease diagnosis based on symptom features; Yang combined PSO and SVM to propose a feature selection method suitable for hyperspectral image classification. The experimental results prove that this hybrid classification method has higher classification accuracy and can extract or select effective bands from multiple bands of hyperspectral images (Yang et al. 2007); Chuang applied the so-called "catfish effect" in the PSO algorithm and used it for feature selection. The experimental results show that the "catfish effect" can help PSO escape from the trap of local convergence, and better results can be obtained compared to other search algorithms (GA, SFS, BSO) (Chuang et al. 2008). In addition, GPSO has been applied multiple times in remote sensing image analysis and processing research, such as image segmentation (Mekhmoukh et al. 2015), image classification (Jiang et al. 2010; Agrawal et al. 2015) and hyperspectral data dimension reduction (Yang et al. 2012). However, feature selection methods based on GPSO are rarely used in high-resolution remote sensing image change detection research.

The role of the fitness function is to evaluate the quality of the generated feature subsets, which serves as the fitness function of the search algorithm to compare with the stop criterion to determine whether to exit the iteration. Different feature selection methods determine different types of fitness functions. The filter's discriminant function measures its quality by analysing the characteristics within the feature subset, regardless of the choice of classifier; the wrapper compares the selected feature subset with the sample set, and the comparison accuracy serves as the standard for measuring the quality of the feature subset. Common filter discriminant functions include correlation, distance, and information gain, while the wrapper discriminant function is the classifier error rate. Which discriminant function to use as the fitness function of the search algorithm should fully consider the purpose of feature selection.

## 4.2 Feature Optimisation Selection Based on GPSO

### *4.2.1 Particle Swarm Algorithm*

To address the issues of local convergence, poor global search capability, and large computational load in heuristic search algorithms, swarm intelligence algorithms based on bionics have emerged. These algorithms are efficient, have strong search capabilities, and can effectively avoid local convergence problems. The Particle Swarm Optimisation (PSO) algorithm seeks the optimal solution through information transmission and sharing among individuals in the group (Eberhart et al. 2001). It is a simple and effective new evolutionary computing technique proposed after genetic algorithms and ant colony algorithms. This algorithm uses an information sharing mechanism, allowing individuals to learn from each other's experiences to promote group development, demonstrating typical swarm intelligence characteristics. The particle swarm algorithm is a type of evolutionary algorithm that starts from a random solution and iteratively finds the optimal solution, evaluating the quality of the solution based on fitness. In the feature selection process of PSO, each random selection can serve as a random solution. The optimal threshold is found through mutual learning among individuals. Compared to other optimisation methods, the PSO optimisation algorithm has a higher degree of automation and stronger algorithm robustness. However, the algorithm has a problem of local convergence, and its overall applicability remains to be discussed. The key to applying the PSO algorithm in feature selection lies in the construction of the fitness function. The standard for constructing the function is based on the degree of object separability of the difference image obtained under this feature selection condition. Generally, if the degree of object separability is high, the difference information is more prominent, which is more conducive to threshold segmentation and extraction of the change area. For most optimisation problems, most optimisation algorithms solve problems by finding the optimal solution in a high-dimensional solution space. In the search process, optimisation algorithms often evaluate individuals through evolutionary operators to help them find the optimal position of the individual. Therefore, the quality of the optimisation result generally depends on the evolution result of the individual. In the PSO algorithm, each individual is considered a particle in a high-dimensional search space, and the combination of these particles is a particle swarm, which represents a series of potential solutions to the optimisation problem. Each particle in the particle swarm represents a potential solution. Like other optimisation problems, the goal of PSO is to find the optimal particle position in the high-dimensional solution space, and this position is evaluated by the fitness function. In this high-dimensional solution space, the positions of the particles are constantly updated. The flight speed and direction of the particles mainly depend on their own flight experience and the flight experience of the surrounding particles. The position of the entire particle swarm is also constantly updated. The change in the position of the group affects the flight of the individual. This process is repeated until the optimal group position is found, which is evaluated by the fitness function. From this process, it can be seen that the construction of the fitness function is the focus of the PSO algorithm. In the PSO algorithm, each particle includes three parts: ① xi represents the current position of the particle; ② vi represents the current speed of the particle; ③ pbesti represents the optimal individual position during the flight of the particle. This position refers to the optimal position the particle has reached, and the fitness value of the particle at this position is the highest. However, this position is constantly updated as the iteration process continues. Generally, the optimal individual position of particle i at time t + 1 is

$$\mathrm{pbest}_i(t+1) = \begin{cases} x_i(t+1), \mathrm{iffitness}(x_i(t+1)) < \mathrm{fitness}\big(\mathrm{pbest}_i(t)\big) \\ \mathrm{pbest}_i(t), \mathrm{iffitness}(x_i(t+1)) \geq \mathrm{fitness}\big(\mathrm{pbest}_i(t)\big) \end{cases} \tag{4.1}$$

In the formula, fitness is the fitness value of the particle at that position. Suppose a particle swarm is composed of n particles, each particle has a D-dimensional position vector xi = (xi1, xi2, …, xiD) and a speed vector vi = (vi1, vi2, …, viD), which determine its flight direction and distance. When the particle moves in the D-dimensional solution space, it will mark an optimal individual position pid = (pi1, pi2, …,…, $p_{iD}$), the entire swarm will also mark a global optimal position $p_{gd} = (p_{g1}, p_{g2}, \ldots, p_{gD})$. Furthermore, each particle will have a fitness value at each moment to evaluate, and this fitness value is calculated by the fitness function previously constructed by the user, which is a quantifiable error value.

Firstly, initialise the parameters and randomly distribute the particles; then find the current optimal particle position, that is, the particle with the highest fitness value, marked as the individual optimal position pbest $_{id}$; then find the global optimal position of the swarm gbest $_{id}$, the formula for updating the velocity and position of the particle at the next moment is as follows:

$$v_{id}(t+1) = \omega v_{id}(t) + c_1 r_1\left(\text{pbest}_{id}(t) - x_{id}(t)\right) + c_2 r_2\left(\text{gbest}_{id}(t) - x_{id}(t)\right) \quad (4.2)$$

$$x_{id}(t+1) = x_{id}(t) + v_{id}(t+1) \quad (4.3)$$

where, $i = 1,2,\ldots, N$; $N$ is the total number of particles in the swarm; $r_1$ and $r_2$ are uniformly distributed random numbers between (0,1); $c_1$ and $c_2$ are learning factors, $c_1$ represents the particle's own learning experience, $c_2$ represents the particle's learning experience from the swarm, generally, the sum of $c_1$ and $c_2$ is considered optimal around 4.0 (Kennedy et al. 1995); $\omega$ is the inertia weight factor, its value is non-negative, it controls the influence of the particle's velocity at the previous moment on the particle at the current moment, the size of the value affects the overall optimisation ability, it is a factor that balances the global and local search capabilities of the swarm (Bourennane et al. 2010). Eberhart believes that if the initial value of $\omega$ factor is set to 0.9, it gradually decreases to 0.4 as the number of iterations increases (Eberhart et al. 2001).

The particle calculates a fitness value based on its position through the optimisation function, each particle has experienced an optimal individual position, the swarm has experienced an optimal swarm position, and the particle updates its own state according to the three principles of its own inertia, optimal individual position, and optimal swarm position.

### 4.2.2 Crossover Operation

To enhance the global search capability of the PSO algorithm and avoid the problem of excessive local convergence, the crossover mutation operation in the genetic algorithm is introduced in the process of updating the particle velocity and position. This enhances its global search capability. The basic principle is that the update change of the offspring particle is not only related to the current particle's position and velocity, but also to the parent particle's velocity and position. Firstly, the algorithm selects particles to participate in the crossover operation from all particles with a certain crossover probability. The specific crossover point is determined by the crossover weight $w_i$, then randomly combines in pairs to perform the crossover operation to produce offspring particles. The crossover process of the position and velocity vectors of the offspring particles is as follows:

$$\text{child}_1(x_i) = w_i \times \text{parent}_1(x_i) + (1 - w_i) \times \text{parent}_2(x_i) \tag{4.4}$$

$$\text{child}_2(x_i) = w_i \times \text{parent}_2(x_i) + (1 - w_i) \times \text{parent}_1(x_i) \tag{4.5}$$

$$\text{child}_1(v_i) = \frac{(\text{parent}_1(v_i) + \text{parent}_2(v_i)) \times |\text{parent}_1(v_i)|}{|\text{parent}_1(v_i) + \text{parent}_2(v_i)|} \tag{4.6}$$

$$\text{child}_2(v_i) = \frac{(\text{parent}_1(v_i) + \text{parent}_2(v_i)) \times |\text{parent}_2(v_i)|}{|\text{parent}_1(v_i) + \text{parent}_2(v_i)|} \tag{4.7}$$

The crossover operation allows the offspring particles to inherit the advantages of the parent particles and maintains the diversity of the particle swarm, enhancing the search capability of the region and escaping from the local optimum. Compared with the basic particle swarm algorithm and the traditional genetic algorithm, the genetic particle swarm optimisation algorithm has a fast search speed and high convergence accuracy (Fig. 4.2).

### 4.2.3 *Fitness Function*

The feature selection process based on GPSO hinges on the setting of the fitness function. The aim is to select features in the feature space that better reflect the changes in the before-and-after temporal images. The discriminant function for feature selection is determined by the differences in the differential image object features. If the image object change detection process is simplified to a binary classification process, a discriminant function suitable for classification can be constructed as the discriminant function for image object feature selection. This book proposes to use the ratio of mean value to variance as the fitness function for GPSO.

According to statistical theory, the mean and variance of data can reflect important information about data distribution. For different object features, if the variance of the feature data is used to standardise the mean of the feature data, and then the differences in the means between different categories are compared. If the means of different categories are clearly distinguishable, it indicates that the feature is important and has the ability to distinguish between two classes of samples; if the means are indistinguishable, the importance of this feature is reduced, therefore the

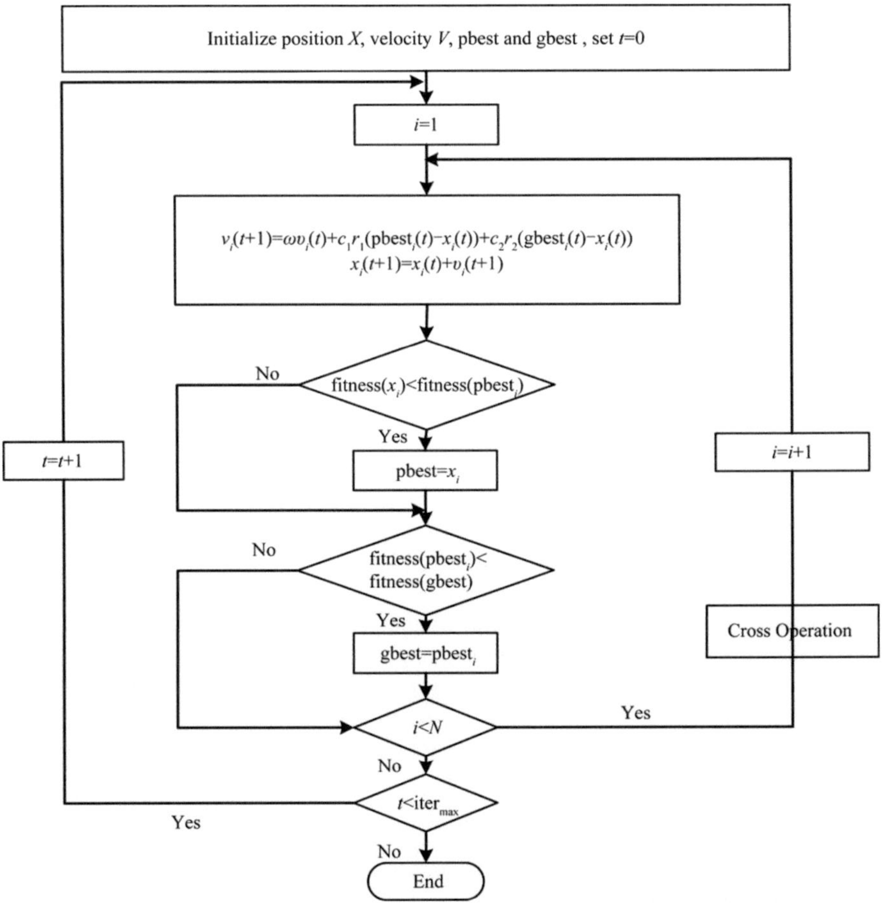


**Fig. 4.2** Flowchart of the GPSO algorithm

ratio of mean value to variance (RMV) of different class features can be used as its fitness function for evaluation.

Let $A$ and $B$ be the sets of values measuring different class features, $A$ is the set of values of a certain feature in the changing class samples (the number of samples is $n_A$), $B$ is the set of values of this feature in the unchanged samples (the number of samples is $n_B$), then the importance of feature $f$ is defined as:

$$S_f = \frac{mean_f(A) - mean_f(B)}{V_f} \tag{4.8}$$

$$V_f = \sqrt{\frac{\mathrm{Var}_f(A)}{n_A} + \frac{+\mathrm{Var}_f(B)}{n_B}} \tag{4.9}$$

In the formula, mean $_f$ $(A)$ and mean $_f$ $(B)$ are the means of sets $A$ and $B$ respectively; $\mathrm{Var}_f$ $(A)$ and $\mathrm{Var}_f$ $(B)$ are the variances of sets $A$ and $B$ respectively. Change detection is to detect the distinguishability of the features of differential image objects, that is, there are only two categories, change and no change, so the mean–variance method is applied to the constructed fitness function.

The features selected for the samples are $M$, and $M$ importance indicators for each category can be obtained, forming the feature importance indicator matrix $S$. From matrix $S$, the average value of the feature importance indicators $S_{\mathrm{AVG}}$ can be calculated:

$$S_{AVG} = \frac{1}{M}\sum_{f=1}^{M} S_f \tag{4.10}$$

From the feature importance matrix $F$, the sum of the feature importance indicators greater than $F_{\mathrm{AVG}}$ is calculated:

$$V_s = \sum_{f=1}^{M} S_f, \quad if\ S_f > S_{AVG} \tag{4.11}$$

The fitness function is constructed as

$$J = V_s \times S_{\mathrm{AVG}}^2 \tag{4.12}$$

Obviously, the larger $S_{\mathrm{AVG}}$ is, the stronger the classification ability of the selected features. The fitness function based on RMV is:

$$\mathrm{Fitness} = -V_s \times \left(\frac{1}{M}\sum_{f=1}^{M} S_f\right)^2 \tag{4.13}$$

### *4.2.4 Multi-feature Optimisation Process Based on GPSO*

Assume $M$ features are selected, then let particle $x_i$ ($M$-dimensional) represent the encoding (position) of the selected features, $v_i$ represents the update speed of the selected feature encoding, the swarm contains $N$ particles, each particle represents a $M$ -dimensional encoding of the selected features (i.e., represents $N$ One of the solutions in the multidimensional solution space). In this multidimensional solution space, particles update their position and velocity through individual and swarm optimisation, i.e., the update of the selected feature encoding.

When the GPSO feature selection algorithm is applied in multi-feature change detection, the most crucial step is how to select the appropriate feature encoding from the feature set. The number of bits in this encoding is the number of features, with each bit corresponding to a feature. The purpose of GPSO feature selection is to search for an optimal set of feature encodings. Figure 4.3 shows the update process of one particle, which is the search process for features. The number of features set in the figure is 6, meaning a single particle contains six features. At the $t$ iteration, the particle position (the encoding of the selected feature) is $x\ (t) = (F1, F2, \ldots, F6)^{\mathrm{T}}$, and the updated velocity calculated by GPSO is $v\ (t) = (v1, v2, \ldots, v6)^{\mathrm{T}}$. At the $t + 1$ iteration, the particle's position is updated to $x\ (t + 1) = (F1', F2', \ldots, F6')^{\mathrm{T}}$. A particle represents one combination of feature encodings, considered a potential solution in the feature selection problem. When the iteration ends, the feature encoding combination included in the swarm optimisation is the final result of feature selection.

The basic process of GPSO feature selection is shown in Fig. 4.4, described as follows:

(1) Parameter initialisation, initial particle number is $N$, learning factors are $c_1$ and $c_2$, inertia weight $\omega$ and maximum iteration number is $\mathrm{iter}_{\max}$;
(2) Assuming the number of features to be selected is $M$, initialise the swarm, each particle $x_{id}$ contains $M$ feature encodings, and the search space is the established object feature space set;
(3) According to the fitness function (4.13), calculate the fitness of each particle, obtaining individual and swarm optimisation $\mathrm{pbest}_i$ and $\mathrm{gbest}_i$;
(4) Update the particle's position and velocity (4.2) and perform crossover operation (4.4)–(4.7);

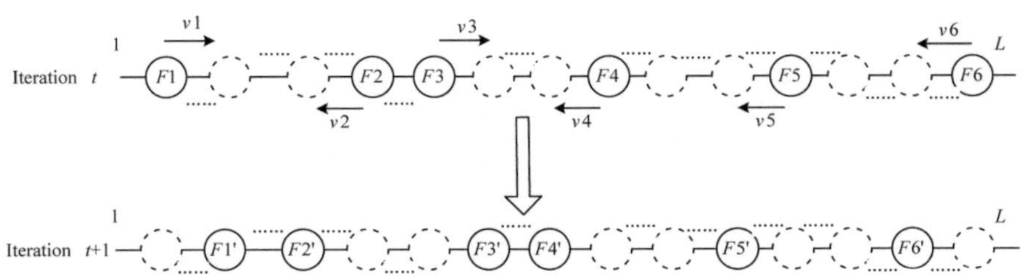


**Fig. 4.3** Single particle update process based on GPSO feature selection algorithm

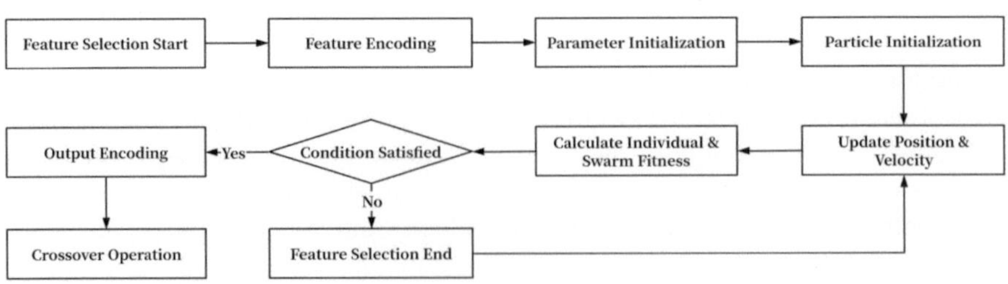


**Fig. 4.4** Basic process of GPSO feature selection algorithm

(5) Until the algorithm reaches the termination condition, exit the operation, otherwise repeat (3);
(6) The final swarm optimisation of the particle is the final solution, and the feature set it contains is the optimal feature subset.

## 4.3 Experiment and Result Analysis

To verify the effectiveness and applicability of the GPSO feature selection algorithm in the multi-feature OBCD process, two sets of high-resolution remote sensing image data were selected for the experiment. The Sequential Float Forward Selection (SFFS) and Cuckoo Search (CS) were compared with GPSO for convergence and detection accuracy.

The CS algorithm is also a swarm intelligence algorithm. The selection of the optimal solution also depends on the comparison of fitness values. However, unlike GPSO, CS updates its position and search path through a "random walk" mechanism. Theoretically, this can have higher global search capabilities than GPSO, and it can also maintain a balance between local search strategies and global space exploration. These two algorithms have similarities and differences with GPSO and can be used as reference algorithms for comparison analysis with GPSO. The SFFS algorithm is a heuristic search algorithm, which is not sensitive to the initial value setting, but it is easy to fall into the problem of local convergence. As a comparison, it can intuitively evaluate the global search capability of GPSO.

### *4.3.1 Experiment Area and Data*

The experimental areas are the Beijing Olympic Park and its surrounding area (Area One) and the farmland area surrounding the Beijing Changping Xiaotangshan Modern Agricultural Demonstration Garden (Area Two). The analysis results from Chap. 3 are directly applied in this experiment to obtain comparable image objects. The experimental data selected are two sets of high-resolution remote sensing image data. The data for Area One are WV-2_2012 and WV-2_2013, and the data for Area Two are FWV-2_2010 and FWV-3_2014. For specific information about the data, refer to Sect. 2.2. Figures 4.5 and 4.6 respectively display the true colour composite images of the two sets of high-resolution remote sensing images for Area One and Area Two. It can be intuitively seen that the main changes in Area One are in the building areas, primarily spectral feature changes. However, due to the interference of building shadows, some false changes may also occur, and it is difficult to identify these changes using only spectral features. The changes in Area Two are in the farmland areas, with changes in both spectral and texture features. These are caused by the seasonal changes and cultivation patterns of crops, and multiple features need to work together to identify them.

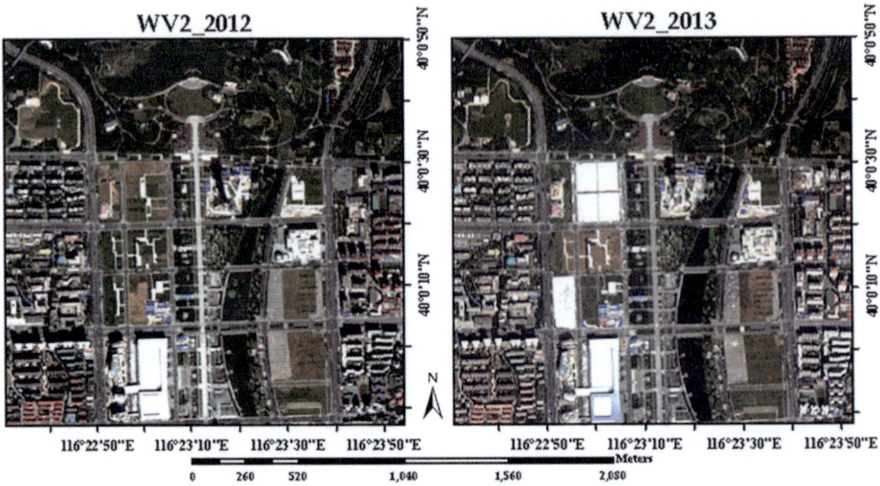


Fig. 4.5 Two images of area one WV-2_2012 and WV-2_2013

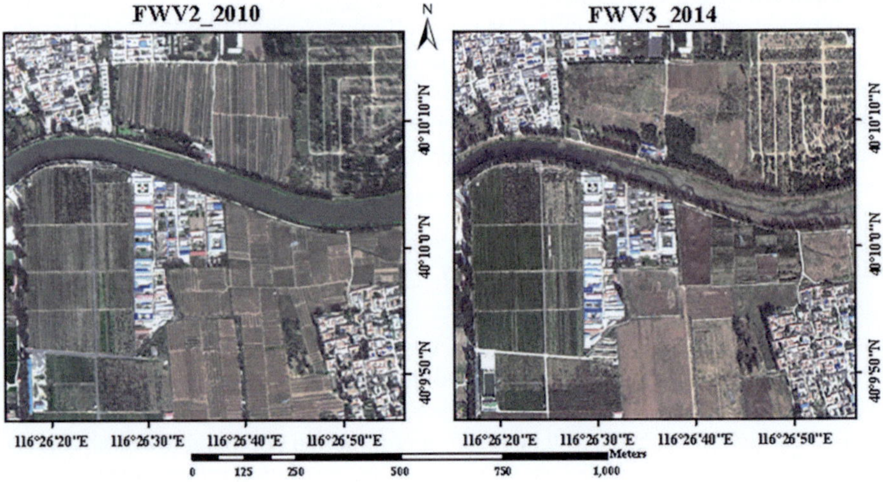


Fig. 4.6 Two images of area two FWV-2_2010 and FWV-3_2014

Data preprocessing is carried out according to the content of Chap. 2, involving geometric precision correction and relative radiometric correction. First, 50 evenly distributed ground control points are selected for geometric precision correction, ensuring that the RMSE is less than one pixel. Then, 50 pseudo-invariant feature points are selected for relative radiometric correction based on robust regression. These operations ensure the comparability of multi-temporal images. For subsequent change detection accuracy verification, 328 and 212 test samples are selected in the two areas respectively as reference data. With the help of visual judgement and field research, the types of changes in the test samples are determined.

**Table 4.1** Spectral, shape, and texture features of image objects

| Area | Spectral features | Shape features | Texture features |
|---|---|---|---|
| Area one | Band Mean | Shape index | GLCM-Correlation |
| | | | GLCM-Contrast |
| | Standard deviation | Density | GLCM-Ang.2nd moment |
| | | | GLCM-Homogeneity |
| Area two | Band mean | Shape index | GLCM-Correlation |
| | | | GLCM-Contrast |
| | NDVI | Density | GLCM-Entropy |
| | | | GLCM- Mean |

### 4.3.2 Feature Extraction of Objects

In the OBIA process, image objects are considered as basic processing units. The generation process of image objects refers to the content of Chap. 3, and suitable image objects are obtained based on the multi-scale segmentation algorithm. To ensure that the two images of the same area have the same boundaries and number of objects, the two image bands are superimposed into one image for MRS segmentation. Section 3.5.2 has discussed the impact of image superposition on the optimisation of segmentation parameters. According to Table 3.11, the optimal segmentation parameters are: for Area One, segmentation scale $S = 166$, shape heterogeneity weight $w_{sh} = 0.6$, compactness heterogeneity weight $w_{cp} = 0.3$; for Area Two, $S = 196$, $w_{sh} = 0.4$, $w_{cp} = 0.4$. As a result, the number of segmented objects in Area One is 826, and in Area Two is 432.

Image objects contain various features. Table 4.1 lists the main image object features of each area. Spectral and shape features are extracted through band and boundary calculations, while texture features are generally calculated using the grey level co-occurrence matrix (GLCM) (Haralick et al. 1973; Honeycutt et al. 2008; Kohli et al. 2013).[1] Each GLCM feature has four directions (0°, 45°, 90°, and 135°), thus forming an image object feature library, which contains a total of 20 features.

### 4.3.3 Convergence Evaluation

The convergence analysis of the GPSO feature selection algorithm can reflect the algorithm's global search capability and convergence speed. By comparing the difference between the average fitness of the particle swarm at each iteration and the global optimal fitness, the convergence level of the algorithm can be intuitively observed. The average fitness $fit_{avg}$ reflects the efficiency of the algorithm. By comparing it with the global optimal fitness $fit_{opt}$, the difference reflects the algorithm's global

[1] The platform for image object feature extraction is eCognation Developer 9.2 (Trimble, Germany).

search capability and convergence level. The closer $fit_{avg}$ is to $fit_{opt}$, the higher the algorithm's global search capability and the better the convergence.

The initial parameter settings of the GPSO feature selection algorithm are shown in Table 4.2. The number of features to be selected in Area 1, $M = 7$, means that each particle contains the encoding of 7 features. To analyse the fitness convergence of the GPSO feature selection algorithm, the SFFS and CS algorithms are selected for comparison, and the iteration process of each algorithm is analysed. In Table 4.3, $Ite_{con}$ refers to the number of iterations when the algorithm converges, and $D_{con}$ refers to the final difference between $fit_{avg}$ and $fit_{opt}$ at convergence. As can be seen from Fig. 4.7a, SFFS converges the fastest, converging at the 35th iteration, but the difference $D_{con}$ between $fit_{avg}$ and $fit_{opt}$ at convergence is large, indicating that the SFFS algorithm has not achieved global search and has fallen into the trap of local convergence; CS converges slowly and is inefficient, but compared to other algorithms, $D_{con}$ is the smallest, the algorithm can avoid local convergence, and has strong global search capability; GPSO reaches convergence at the 45th iteration, $D_{con}$ is slightly larger than CS but the difference is small, it has basically jumped out of the trap of local convergence, and the convergence curve of GPSO is more stable than other algorithms, and the efficiency of GPSO algorithm in finding the optimal value is higher. Figure 4.7b shows the convergence process of each algorithm in Area 2, and the analysis results of each algorithm in Area 2 are similar to those in Area 1.

**Table 4.2** Initial parameter settings of the GPSO feature selection algorithm[a]

| Initial parameters | Initial population size $N$ | Learning factor $c_1$ | Learning factor $c_2$ | Inertia weight $\omega$ | Maximum iteration number $iter_{max}$ | Number of features to be selected $M$ |
|---|---|---|---|---|---|---|
| Area 1 | 80 | 2.6 | 1.6 | 0.9 | 100 | 7 |
| Area 2 | 60 | 2.8 | 1.3 | 0.9 | 80 | 6 |

[a] The GPSO algorithm was implemented using Matlab (2010b)

**Table 4.3** Results of fitness convergence analysis

| Region | Algorithm | $Ite_{con}$ | Final $fit_{avg}$ | $D_{con}$ |
|---|---|---|---|---|
| Region one | SFFS | 35 | – 327 | 88 |
| | CS | 75 | – 402 | 13 |
| | GPSO | 45 | – 388 | 27 |
| Region two | SFFS | 29 | – 296 | 66 |
| | CS | 60 | – 355 | 7 |
| | GPSO | 39 | – 345 | 17 |

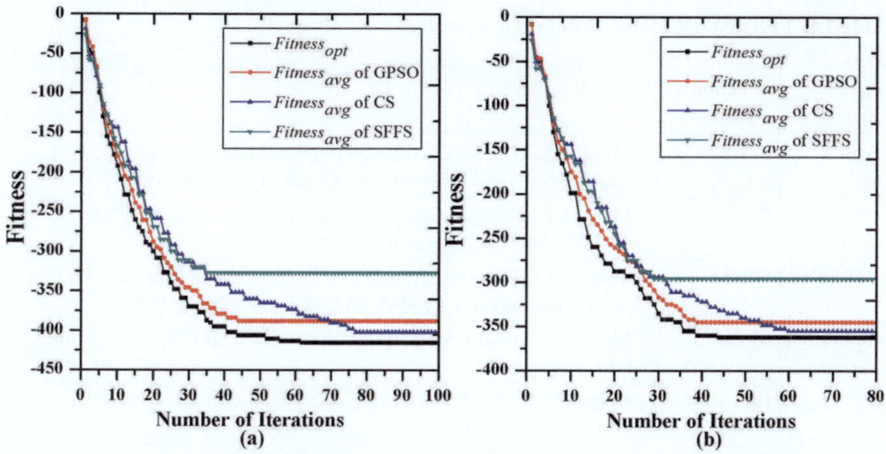


**Fig. 4.7** Comparison of fitness convergence analysis for GPSO, CS and SFFS algorithms

### *4.3.4 Evaluation of Change Detection Accuracy*

Change detection in image objects requires the selection of optimal object features or feature combinations. Therefore, it is necessary to construct a difference image, which concentrates the change information of multi-temporal images on a single image. The feature information of this image object is the feature difference information of multi-temporal image objects. Then, change areas are extracted from the difference image through threshold segmentation and other methods to complete change detection. In object-oriented change detection, the methods commonly used to construct difference images are object direct comparison and image transformation, each with its own advantages and disadvantages. Considering that it is a multivariate (multi-feature) transformation detection problem, the Multivariate Alternative Detection (MAD) algorithm based on canonical correlation analysis (Nielsen et al. 1998) is used to construct difference image objects. Then, the Histogram Curvature Analysis (HCA) is used to set the threshold and extract the change area. This change detection algorithm is called the OB-HMAD (object based-hybrid MAD) algorithm. The OB-HMAD method has unique advantages in multi-channel image processing and can maximise the retention of change information in image objects. It highlights change information through non-linear transformation to obtain difference images. For the mathematical theory details of the OB-HMAD method, see Sect. 5.5.1.[2]

To verify the applicability of the GPSO feature selection algorithm in OBCD, the selected feature combinations are input into the OB-HMAD model to obtain the corresponding change detection results, and the accuracy of the change detection results is evaluated. Combined with reference data, the accuracy is evaluated using the error confusion matrix method, and the accuracy of OBCD is measured using indicators such as the false negative rate (FNR), false positive rate (FPR), overall accuracy (OA), and Kappa coefficient (Table 4.4).

[2] The OB-HMAD method was programmed and implemented on the Matlab 2010b platform.

**Table 4.4** Feature selection results of each algorithm

| Region | Feature selection algorithm | Feature selection results |
|---|---|---|
| Region one | SFFS | Mean, Std, GLCM-Correlation (90°), GLCM-Contrast (0°, 90°), GLCM-Ang. 2nd moment (0°), GLCM-Homogeneity (0°) |
| | CS | Mean, GLCM-Correlation (0°, 90°), GLCM-Contrast (0°, 90°), GLCM- Ang. 2nd moment (0°, 45°) |
| | GPSO | Mean, GLCM-Correlation (0°, 90°), GLCM-Contrast (0°), GLCM-Ang. 2nd moment (0°, 45°), GLCM-Homogeneity (0°) |
| Region two | SFFS | Mean, NDVI, GLCM-Correlation (0°, 90°), GLCM-Contrast (0°), GLCM-Mean (0°) |
| | CS | Mean, NDVI, GLCM-Correlation (0°), GLCM-Contrast (0°, 90°), GLCM-Entropy (0°) |
| | GPSO | Mean, NDVI, GLCM-Correlation (0°), GLCM-Contrast (0°, 90°), GLCM-Entropy (0°) |

*Note* Mean refers to the band mean; Std refers to the standard deviation

There are seven figures in Fig. 4.8, which are used to analyse the issues of omission and commission errors. Four sample areas were selected, where Area 1 is the change area, mainly involving omission errors; Areas 2–4 are unchanged areas, mainly involving commission errors. Figure (d) differs significantly from the other figures, and compared to Figure (c), it has more omission and commission errors; Figure (g) has the highest agreement with Figure (c), but it also has omission and commission errors. Comparative analysis of the four sample areas shows that Area 1 is considered a change area in Figure (c), but is omitted in Figures (d) and (e); Area 3 is an unchanged area in Figure (c), but is misidentified as a change area in Figures (d) to (f) due to spectral errors caused by different solar incidence angles, while Figure (g) correctly identifies Area 3 as an unchanged area, which is due to the influence of the GLCM-Homogeneity feature; The commission error in Area 2 is due to the shadow of high-rise buildings, and Area 4 is a pseudo-change area caused by the incidence angle offset of the remote sensing sensor, but all algorithms fail to identify this pseudo-change area.

Similar to Region One, Fig. 4.9 also lists four types of sample areas to analyse the omission and commission errors within the area. Areas 1 and 2 are change areas, mainly involving omission errors; Areas 3 and 4 are unchanged areas, mainly involving commission errors. Area 1 is a farmland area, and its changes are mainly due to the changes in the farmland ridge pattern, which are texture feature changes. Figure (d) is a single spectral feature that fails to identify this change; Similar to Area 1, Area 2 also belongs to texture changes, mainly changes in the direction of the ridges, so Figures (d) and (e) fail to identify; Area 3 is a spectral change, but this is a change in the growth of crops, so Figure (c) considers this as an unchanged area, affected by spectral features, Figures (d), (f) and (g) all misidentify it as a change area; A similar commission error also exists in Area 4. It can be seen that although multi-features can identify many texture or shape feature change areas that single features cannot identify, they will also be affected by the interference of other

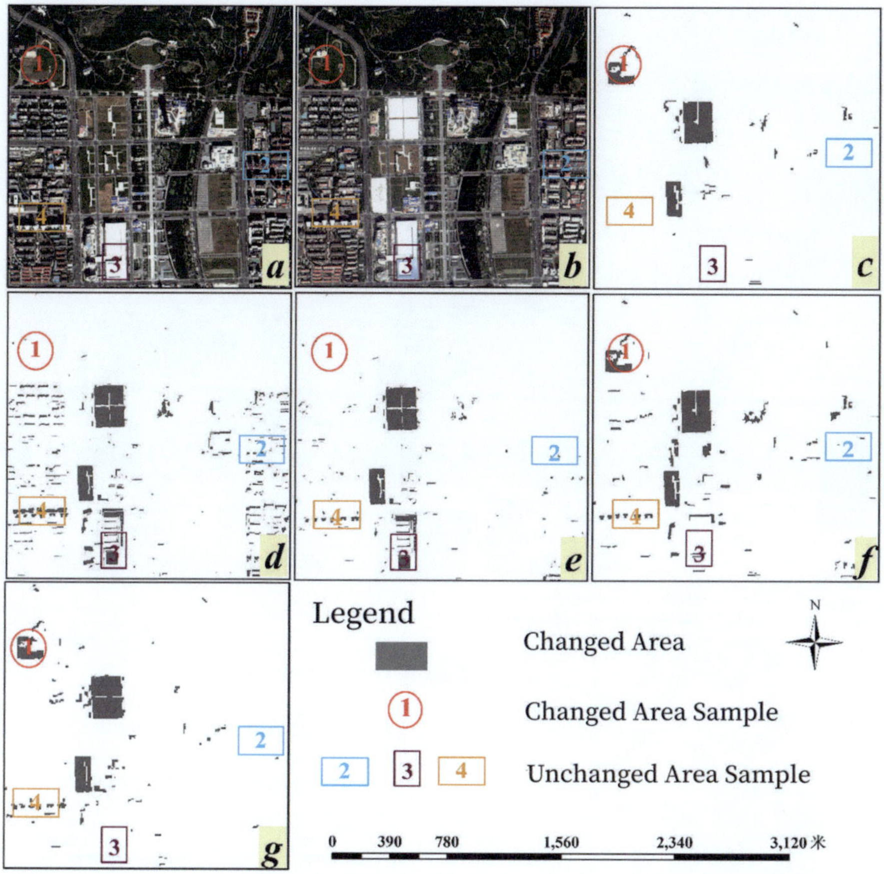


**Fig. 4.8** OB-HMAD change detection results based on various feature selection algorithms for region one **a** WV2_2012; **b** WV3_2013; **c** reference results based on ground truth data; **d** OBCD results based on single feature (band mean); **e** multi-feature OBCD results based on SFFS feature selection; **f** multi-feature OBCD results based on CS feature selection; **g** multi-feature OBCD results based on GPSO feature selection

features, thereby affecting the final accuracy of OBCD. Therefore, when inputting features in multi-feature OBCD, the main change features of the area should be established to avoid this interference.

Table 4.5 shows the results of the OBCD accuracy evaluation using different feature selection methods. It can be seen that the omission and commission rates of single feature OBCD are higher than other algorithms, and the overall accuracy and Kappa are also lower than other algorithms, and it cannot identify more types of changes; Among other structures based on multi-feature OBCD, the SFFS algorithm has a higher omission rate (27.23 and 54.35%), and the evaluation results of CS and GPSO are closer, with a slightly higher commission rate than GPSO (22.47 and 26.49%); GPSO has a higher overall accuracy (83.59 and 84.17%) and Kappa coefficient (0.6314 and 0.6771).

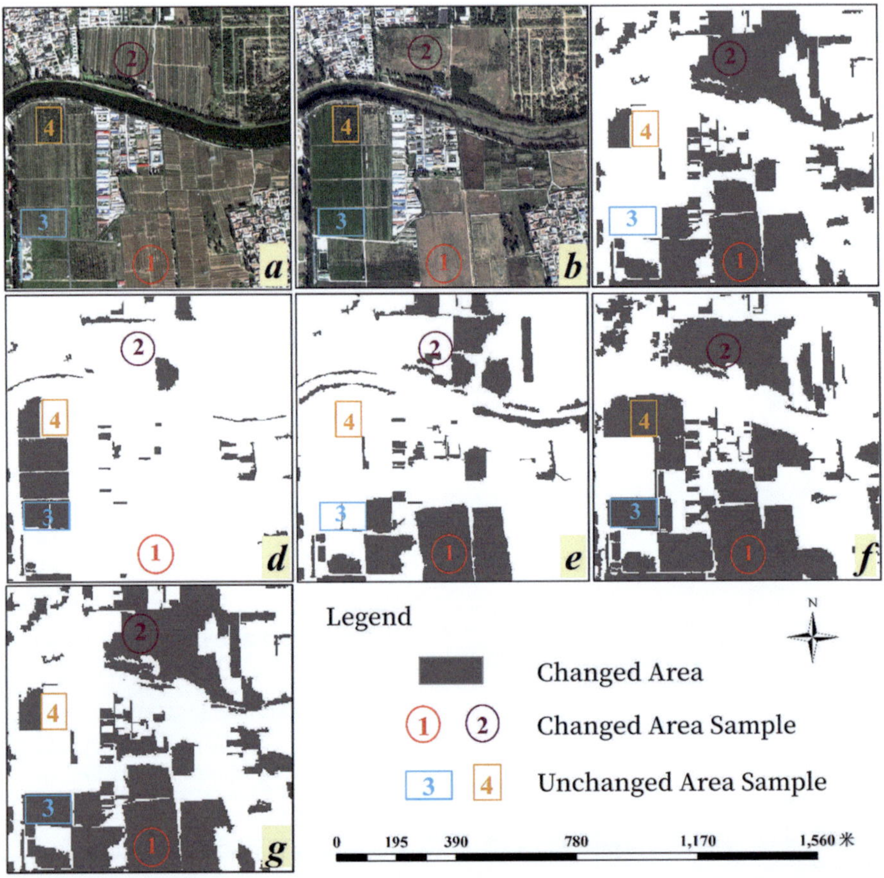


**Fig. 4.9** OB-HMAD change detection results based on various feature selection algorithms for region two **a** FWV2_2010; **b** FWV3_2014; **c** reference results based on ground truth data; **d** OBCD results based on single feature (band mean); **e** multi-feature OBCD results based on SFFS feature selection; **f** multi-feature OBCD results based on CS feature selection; **g** multi-feature OBCD results based on GPSO feature selection

## 4.4 Discussion

When the GPSO feature selection algorithm is applied in the process of multi-feature OBCD of VHR remote sensing images, the setting of some parameters or functions will affect the accuracy and efficiency of the algorithm itself. In this experiment, a set of VHR images FWV-2_2010 and FWV-3_2014 from Region Two were selected as experimental data. Through sensitivity analysis, we explored the impact of the number of candidate features $M$ and the initial population size $N$ on the accuracy and efficiency of the algorithm, and compared the OBCD accuracy changes of the GPSO feature selection algorithm under different fitness functions.

**Table 4.5** Accuracy evaluation of OBCD results based on different feature selection methods

| Region | Method | Miss rate (%) | False alarm rate (%) | Overall accuracy (%) | Kappa |
|---|---|---|---|---|---|
| Region one | Single feature | 32.35 | 57.45 | 49.22 | 0.0726 |
| | SFFS | 27.23 | 16.47 | 72.63 | 0.4517 |
| | CS | 11.76 | 22.47 | 79.91 | 0.5712 |
| | GPSO | 8.82 | 19.16 | 83.59 | 0.6314 |
| Region two | Single feature | 69.56 | 32.43 | 53.33 | 0.0648 |
| | SFFS | 54.35 | 14.87 | 70.06 | 0.3267 |
| | CS | 6.52 | 26.49 | 78.00 | 0.5187 |
| | GPSO | 10.87 | 18.92 | 84.17 | 0.6771 |

### *4.4.1 Impact of Candidate Feature Number*

As an initial parameter of the GPSO feature selection algorithm, the size of the candidate feature number $M$ will affect the accuracy of the algorithm. A larger $M$ makes the features of OBCD more abundant, containing more change information, but it may also cause data redundancy; a smaller $M$ can improve the efficiency of the algorithm, but it may lose key change information. Therefore, it is necessary to quantitatively analyse the sensitivity of the candidate feature number to the accuracy and efficiency of the algorithm.

Figure 4.10 shows the relationship between the candidate feature number $M$ and the overall accuracy of OBCD and the efficiency of the algorithm. The algorithm's computing platform is an Intel Core i7 2.93-GHZ CPU with 4 GB of memory. The running time of the algorithm is used as the standard for evaluating the efficiency of the algorithm (right $Y$ axis), and the overall accuracy of OBCD is calculated based on the OB-HMAD algorithm and the error confusion matrix (left $Y$ axis). The settings of other parameters of the algorithm are initial population size $N = 60$, learning factor $c_1 = 2.8$ and $c_2 = 1.3$, inertia weight $\omega = 0.9$, maximum iteration times iter $_{\max}$ The feature library has 20 features, and the selection range for the number of features to be selected, denoted as $M$, is [1:20], with a step size of 1. This gives us the sensitivity curve of the selected features $M$ with respect to overall accuracy (OA) and runtime (time). As can be seen from Fig. 4.10, for the change in runtime, the time of the SFFS algorithm rapidly increases with the increase of $M$, indicating that the choice of $M$ has a significant impact on the SFFS algorithm. For the GPSO algorithm, when $M$ is less than 10, the time does not change significantly with the change of $M$, but when $M$ is greater than or equal to 10, the time rapidly increases with the increase of $M$. The time change of the CS algorithm is similar to that of GPSO, and the sensitivity of $M$ to time is consistent for both algorithms.

For the overall accuracy change of OBCD, as can be seen from Fig. 4.10, the OA change curves of each algorithm are slightly different. By performing statistical analysis on the change of OA, we obtain the statistical results as shown in Table 4.6. Combining the analysis of Fig. 4.10 and Table 4.6, it can be seen that the OA of the SFFS algorithm reaches its peak when $M$ is 7, the OA of the CS algorithm

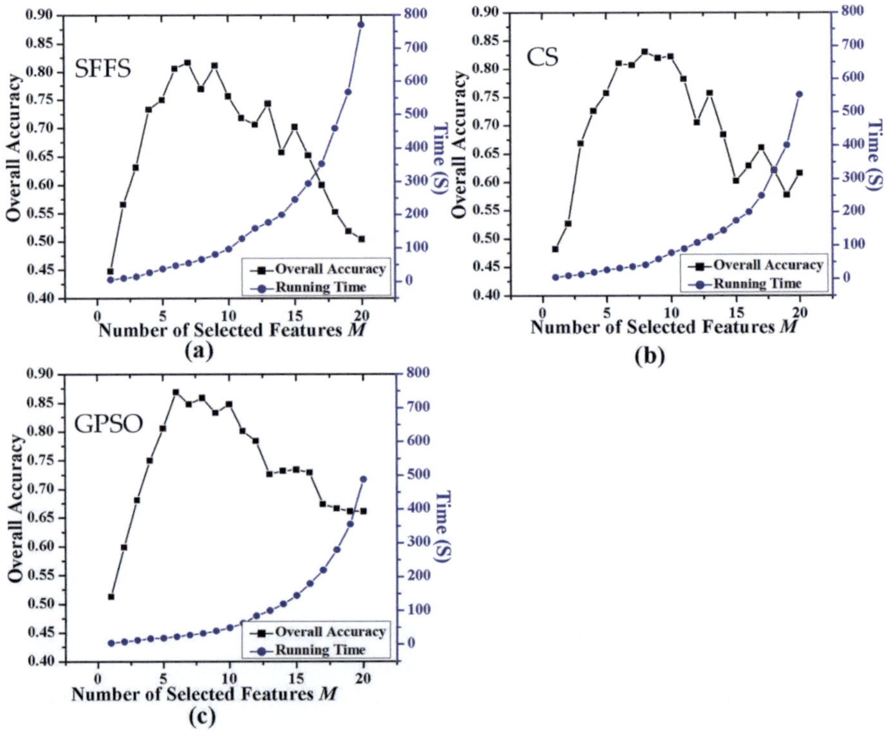


**Fig. 4.10** The impact of the selection of the number of features to be selected in each feature selection algorithm on the accuracy and runtime of OBCD

reaches its peak when $M$ is 8, and the OA of the GPSO algorithm is highest when $M$ is 6. Moreover, the peak and average values of GPSO are relatively higher than other algorithms, reaching 86.9% and 73.1% respectively. Analysis of the standard deviation and average change rate of OA shows that the change in OA of GPSO is smaller than that of other algorithms (9.1% and 5.3%), and the OA change curve of GPSO is smoother, indicating that the sensitivity of GPSO's OA to $M$ is lower and less likely to be affected. It can be seen that when $6 \leq M \leq 10$, the accuracy of OBCD is higher and the runtime cost is lower.

**Table 4.6** Statistical results of the sensitivity analysis of the number of features to be selected on the overall accuracy of OBCD

| Feature selection algorithm | Maximum OA (%) | Maximum OA corresponding to $M$ | Average OA (%) | Standard deviation of OA (%) | Average change rate of OA (%) |
|---|---|---|---|---|---|
| SFFS | 81.7 | 7 | 67.0 | 10.9 | 8.2 |
| CS | 83.1 | 8 | 69.4 | 10.2 | 7.8 |
| GPSO | 86.9 | 6 | 73.1 | 9.1 | 5.3 |

### 4.4.2 The Impact of the Initial Population Size in GPSO

The initial population size $N$ refers to the number of particles in the particle swarm, representing the number of initial solutions, and is also one of the initial parameters of GPSO. Its selection will to some extent affect the convergence and runtime of the GPSO algorithm. Generally speaking, a smaller population size will save runtime cost but reduce global search capability, while a larger population size can enhance global search capability but bring higher runtime cost. Trelea believes that a suitable selection range for the initial population size of GPSO is [20:100] (Trelea 2003). In this experiment, the values of $N$ were selected as 20, 40, 60, 80, and 100, denoted as 20GPSO, 40GPSO, 60GPSO, 80GPSO, and 100GPSO. Other parameters were set as the number of features to be selected $M = 6$, learning factor $c_1 = 2.8$ and $c_2 = 1.3$, inertia weight $\omega = 0.9$, with a maximum iteration count of $iter_{max} = 80$. Therefore, the sensitivity curves of fitness convergence for different population sizes are shown in Fig. 4.11.

As can be seen from Fig. 4.11, the convergence curves of the GPSO for these five population sizes are slightly different, but the overall trend is consistent, converging around 30–40 iterations, and there are differences in the final fitness values at convergence. As shown in Table 4.7, it can be seen that 20GPSO converges the fastest, reaching convergence at the 29th iteration, but the final average fitness is higher than other algorithms, indicating a problem of falling into local convergence; the final fitness values of 60GPSO and 80GPSO are relatively close, but the running cost of 80GPSO is almost twice that of 60GPSO; the final fitness value of 100GPSO is slightly higher than 60GPSO and 80GPSO, showing that a larger population size

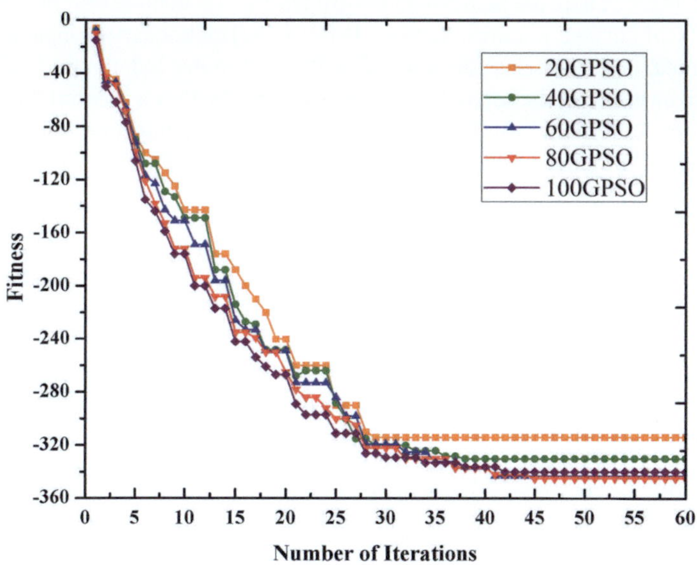


**Fig. 4.11** Impact of different population sizes on fitness convergence in the GPSO feature selection algorithm

**Table 4.7** Sensitivity analysis results of initial population size

| Algorithm | Iteration count at convergence | Final $fit_{avg}$ | Running time/s |
|---|---|---|---|
| 20GPSO | 29 | – 314 | 30 |
| 40GPSO | 38 | – 330 | 45 |
| 60GPSO | 41 | – 343 | 70 |
| 80GPSO | 45 | – 345 | 135 |
| 100GPSO | 46 | – 340 | 180 |

does not necessarily result in higher global search capability. From the above, it can be seen that the initial population size can affect the convergence of GPSO, but the more appropriate initial population size should be 60–80.

### 4.4.3 Comparison of Other Fitness Functions

The fitness function is a standard for evaluating the applicability of an algorithm, and it is necessary to explore the impact of different fitness functions on the convergence of the algorithm. The JMD and NNC fitness functions are chosen as references to compare the convergence of the RMV fitness function, and are respectively denoted as GPSO-RMV, GPSO-JMD, and GPSO-NNC.

1. Fitness function based on JMD

Distance-based fitness functions generally rely on the separability of samples calculated by distance. There are many distance discrimination methods to characterise the separability of surface features, such as JMD, BD (Bhattacharyya distance), intra-class distance, and inter-class distance. Compared to other indicators, JMD is more suitable for expressing the separability of categories. JMD is a spectral feature separability index, commonly used in feature selection for hyperspectral remote sensing bands (Chang-Yao et al. 2006; Eismann 2012). The formula for JMD is as follows (Eismann 2012):

$$J_{ij} = \left\{ \int_x \left[ \sqrt{p(X/\omega_i)} - \sqrt{p(X/\omega_j)} \right]^2 dX \right\}^{1/2} \tag{4.14}$$

In the formula, $p(X/\omega_i)$ is the conditional probability density, i.e., the probability that the i-th pixel belongs to the $\omega_i$-th category; the value range of $J_{ij}$ is [0, 2], and its size represents the degree of separability between samples.

2. Fitness function based on NNC

The nearest neighbour method is a non-parametric pattern recognition method, using its classification error rate as the feature evaluation function. Suppose there is a set of n samples X, each sample has been marked with a category. Suppose there are c pattern samples, each category has $n_i$ samples, $i = 1, 2, \ldots, c$, the discriminant function of the nearest neighbour method is

$$d_i(X) = -\min\|X - X_{ik}\|(k = 1, 2, \ldots, n_i) \tag{4.15}$$

If there is $d_i(X) > d_j(X), i \neq j$, then $X$ is classified into the $i$ th category. The steps to construct the fitness function using NNC are as follows: ① Randomly divide the samples into training samples and test samples; ② Reconstruct the training samples and test samples using each selected feature combination, apply NNC to evaluate the test samples, and obtain the accuracy rate $R$; ③ Define the number of selected features $M$, then the fitness function is constructed as

$$J = R^{(1+M^{0.5})} \tag{4.16}$$

It is known that the smaller $M$ is, the higher the accuracy rate is, and the larger the fitness $J$ value is, which is in line with the principle of feature selection.

3. Comparative Analysis of Fitness Function Convergence

Experiments were conducted on the GPSO-RMV, GPSO-JMD and GPSO-NNC algorithms, with all other parameters of the algorithms set identically. Figure 4.12 shows the convergence analysis of GPSO based on three fitness functions. The red line is the convergence curve of the average fitness, and the black line is the convergence curve of the best fitness of the population. The closer the two curves are, the better the global convergence of the algorithm. The average fitness $\text{fit}_{\text{avg}}$ and the best fitness of the population fit $_{\text{opt}}$ maintain a relatively close distance. In the first 20 iterations, the convergence speed is fast and the two curves coincide very well. Relatively speaking, the convergence speed of GPSO-JMD is slightly faster, and $\text{fit}_{\text{avg}}$ and $\text{fit}_{\text{opt}}$ converge at 35 and 38 iterations respectively, but there is a large gap between the final convergence values of the two, indicating that the algorithm has a problem of local convergence; The gap between the $\text{fit}_{\text{avg}}$ and $\text{fit}_{\text{opt}}$ of the GPSO-NNC algorithm is consistently small, but it gradually converges until 50 iterations; The $\text{fit}_{\text{avg}}$ and $\text{fit}_{\text{opt}}$ of the GPSO-RMV algorithm converge almost simultaneously at 40 iterations, and the gap between the two is small. Considering the reliability and running cost of the algorithm, RMV is still the most suitable as the fitness function of the GPSO algorithm.

## 4.5 Summary

This book applies the genetic particle swarm optimisation algorithm to the multi-feature selection process of OBCD, achieving feature optimisation selection based on GPSO. The main research content and conclusions are as follows.

(1) A multi-feature optimisation selection process based on GPSO is proposed. The genetic particle swarm optimisation algorithm is introduced into the research of OBCD multi-feature selection. Based on the GPSO algorithm, the ratio of mean to variance (RMV) is selected as the fitness evaluation function, and a multi-feature optimisation process based on GPSO is designed on this basis.

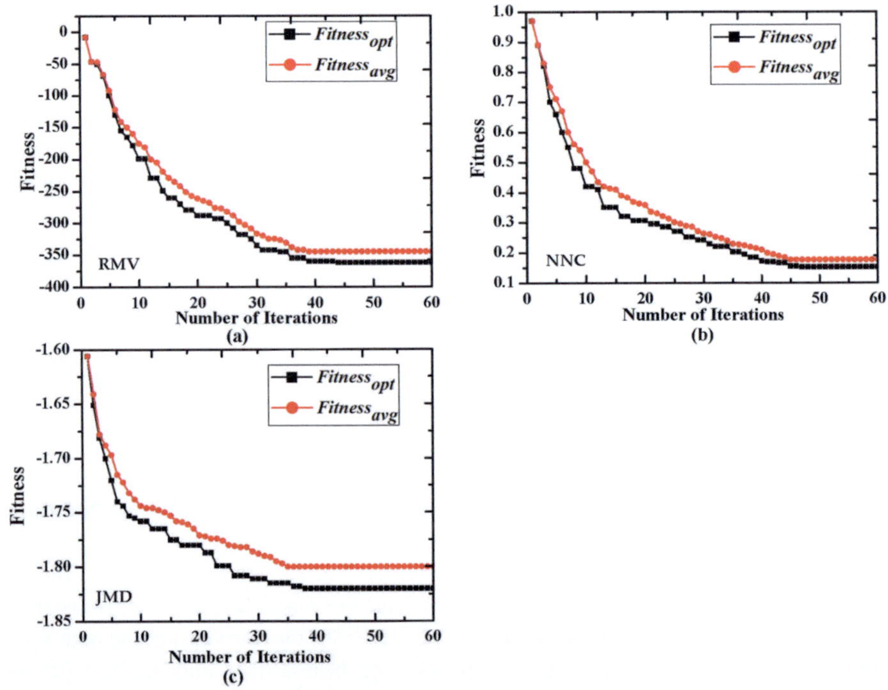


**Fig. 4.12** Comparison of convergence of different fitness functions (colour figure attached later)

(2) The convergence of the GPSO feature optimisation selection algorithm is compared and analysed, and its applicability and reliability are verified through the accuracy evaluation of OBCD. The GPSO feature selection algorithm is applied to the multi-feature OBCD experiment with SFFS and CS algorithms as controls. The feature selection algorithm based on GPSO can avoid the problem of local convergence and has a faster convergence speed; In the OBCD experiment based on OB-HMAD, the multi-feature OBCD result is significantly better than the single feature result, and the OBCD result based on GPSO feature selection has a lower false negative rate and false positive rate, as well as a higher overall accuracy (83.59 and 84.17%), and the selected feature combination has more change information, which can identify more types of changes.

(3) The sensitivity analysis results of the initial parameters of the GPSO feature optimisation selection algorithm show that the number of candidate features $M$ and the initial population size of the particle swarm $N$ will have a certain impact on the accuracy, convergence and efficiency of the algorithm. The accuracy of the GPSO feature selection algorithm is relatively insensitive to $M$, with relatively stable changes, but the running time will increase rapidly. When $6 \leq M \leq 10$, the algorithm has high accuracy and the time cost is not too high; The convergence speed and other convergence indicators of the algorithm have a strong sensitivity to $N$, and the experiment proves that when $N = 60$, the convergence speed and final fitness value of the algorithm are relatively ideal. Compared with JMD and NNC, RMV has advantages in algorithm convergence and reliability, and is more suitable as the fitness function of GPSO.

# Chapter 5
# Multi-feature Change Detection of Image Objects

Addressing the issue that traditional change detection methods cannot or are difficult to identify types of changes, this book improves the change vector analysis method. According to the dominant features of each image object, different adaptive weights are assigned to each feature, constructing a multi-feature change detection theoretical framework based on adaptive weight change vector analysis. This framework mainly includes: ① calculation methods for change intensity and change direction; ② setting methods for adaptive weights; ③ expression of the algorithm in the polar coordinate system. Under this framework, a "two-step" detection pattern for multi-feature changes of high-resolution image objects has been developed. Firstly, the EM algorithm is used on the difference image to obtain the threshold of change intensity, achieving the extraction of changing objects; then, for these objects, the *K*-means method is used to distinguish different change directions, achieving the identification and differentiation of different types of changes.

Multi-feature object-oriented change detection plays an increasingly important role in the field of multi-temporal analysis and application research of high-resolution remote sensing images (Blaschke 2010; Blaschke et al. 2014). Traditional single-feature change detection cannot detect multiple types of changes and cannot reflect the advantages of multi-feature analysis in OBIA. As can be seen from the analysis in Chap. 4, single-feature change detection has many omissions and false detections, and the feature combination problem in multi-feature change detection will also affect the accuracy and efficiency of change detection.

There are many conventional change detection methods, which have been highlighted in Chap. 1. Among them, change vector analysis (CVA) is commonly used. With the widespread application of multispectral and high-resolution images, these change detection methods also need to be extended to higher-dimensional feature spaces. For example, for CVA, the change intensity can still be calculated in the original way, but it is difficult to describe the change direction in the multi-dimensional feature space. Multiple direction cosine values can be used to form an output value, but this will lose the direction of the vector (Xian et al. 2010). In response to how to

 
Q. Chen et al., *Multiple Feature Object-Based Change Detection Technology for Very High-Resolution Remote Sensing Image*, https://doi.org/10.1007/978-981-95-3917-8_5

express the CVA change direction and other change information in the $n$-dimensional feature space, many scholars have proposed corresponding solutions. Nackarets et al. proposed an improved CVA technique, using the Cartesian coordinate system and clustering algorithm to identify the various directions of the change vector (Nackaerts et al. 2005); Johnson et al. used a "sector-coding" method to solve the problem of multi-temporal image overlap with CVA (Johnson et al. 1998); Allen and Kupfer used the direction cosine to calculate the direction of the change vector, which is currently a widely used multi-dimensional CVA method (Allen et al. 2000); Carvalho proposed using spectral correlation mapping to measure the direction of the change vector, and using Mahalanobis distance to measure change intensity (Carvalho Júnior et al. 2011); Bovolo proposed a polar coordinate system method to describe the change direction in the multi-dimensional feature space (Bovolo et al. 2007). Most studies simply simplify or convert the input feature items of multi-dimensional CVA to dual-band, thereby calculating the change direction, but this simplification or conversion method mostly requires the assistance of prior knowledge, which will cause the omission or distortion of multi-feature change information. In addition, in the process of solving the change intensity and change direction in multi-feature CVA, a unified standardization method will be used for each feature, thereby standardizing the values to a unified number field (Bovolo et al. 2008). This standardization method does not consider the different contributions of each feature in different types of changes. Using a unified weight to standardize will lose feature diversity and does not consider the role of the dominant features of each image object.

In view of the above problems, this book constructs a theoretical framework for self-adaptive weight change vector analysis (SAW-CVA). It assigns an adaptive weight to each feature according to different change vectors and dominant features, highlighting the diversity of each feature and the main contribution of the dominant feature in the OBCD process, and preserving the change information in the $n$-dimensional feature space to the greatest extent, in order to identify different types of changes. A multi-feature OBCD process based on the SAW-CVA method is constructed, and a comparative analysis with the standard CVA is conducted through experiments. In addition, a comparative analysis of MAD and other change detection methods suitable for multi-feature OBCD is also conducted, and a comparative discussion on the threshold segmentation method is carried out. The framework of the research content of this chapter is shown in Fig. 5.1.

This chapter includes six parts. Section 5.1 Standard Change Vector Analysis; Sect. 5.2 Theoretical Framework for Multi-feature Change Detection Based on SAW-CVA; Sect. 5.3 Multi-feature Change Detection of Image Objects Based on SAW-CVA; Sect. 5.4 Experiments and Result Analysis; Sect. 5.5 Summary and Discussion; Sect. 5.6 Summary of this Chapter.

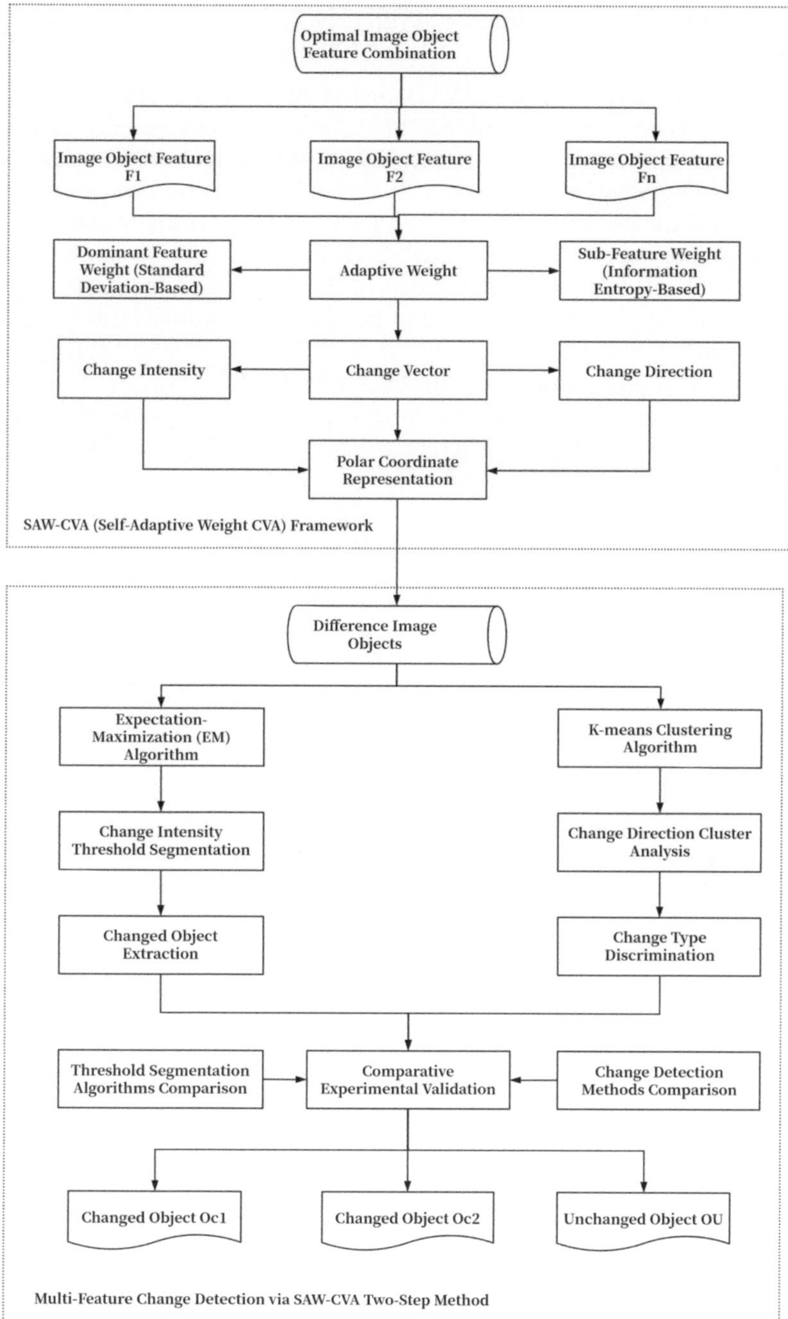


**Fig. 5.1** Framework of multi-feature change detection research based on SAW-CVA image objects

## 5.1 Standard Change Vector Analysis

Standard Change Vector Analysis (CVA) based on the spectrum is used to analyse the spectral differences of images at different times through a change vector described by both change direction and change intensity. In fact, the features of image objects are described by one-dimensional column vectors of each band of the image. By performing difference operations on two remote sensing images, the change value of each pixel or object is obtained, which serves as the change vector. The modulus of the change vector is usually calculated by the Euclidean distance, which serves as the change intensity, while the direction of the change vector reflects the content or type of change (Bovolo et al. 2008).

If there are two geometrically corrected multispectral remote sensing images, taken at times $t_1$ and $t_2$ covering the same area, denoted as images $X^{t1}$ and $X^{t2}$, where $b$ represents their band number, then $b = 1, 2, \ldots, B$. Most change detection techniques identify changes by comparing the differences in corresponding pixels or objects in each band of the two images (Singh 1989). In Change Vector Analysis (CVA), this comparison algorithm is implemented by applying change vectors to extract spectral change information from $n$-dimensional feature space. Therefore, subtracting the spectral feature column vectors of the two images yields a difference image.

$$\Delta X = X^{t2} - X^{t1} \tag{5.1}$$

In the equation, $\Delta X$ represents a pixel or object of a difference image containing change information. If the images being compared are multispectral, then $\Delta X$ is composed of $B \Delta X_b$ components, where $\Delta X_b$ represents the $b$th component of $\Delta X$. The change intensity $\rho$ of a pixel or object is given by

$$\rho = \sqrt{\sum_{b=1}^{B} (\Delta X_b)^2} \tag{5.2}$$

This formula shows the spectral difference of multispectral images at two time points. The larger the $\rho$, the greater the probability of change in the image pixel or object. Therefore, threshold segmentation can be used to distinguish between changed and unchanged pixels or objects in the difference image. Generally, if the input feature of CVA is a single feature, the change intensity can be calculated using Eq. (5.2). However, if the input feature of CVA is no longer a single spectral feature, these features need to be numerically standardised, otherwise the different ranges of each feature cannot be directly compared or merged. Regardless of the standardisation method, the different contributions of different features to change detection and the differences in the dominant features of each object are ignored, resulting in overly concentrated change vector directions, which is not conducive to the identification of change directions.

If it is a two-dimensional change detection problem, the change direction can be calculated using Eq. (5.3). However, if it is a multi-dimensional change detection problem, all change directions cannot be calculated. The conventional approach is to transform the multi-dimensional problem into a two-dimensional problem through dimension reduction (Bruzzone et al. 2000b; Celik et al. 2010; Prieto et al. 2000), such as using principal component transformation to transform each band (Deng et al. 2008), or by pairwise comparison (Chen et al. 2003). Equation (5.3) is a method for solving the change direction in two-dimensional space, but this method generally requires the assistance of prior knowledge and cannot ensure that the change information can be fully reflected in this two-dimensional space. Similarly, the dimension reduction method does not fully consider the differences between features.

$$\theta = \arctan\left(\frac{\Delta X_m}{\Delta X_n}\right), \left(-\frac{\pi}{2} \leq \theta \leq \frac{\pi}{2}\right) \tag{5.3}$$

In view of the above problems with standard CVA, a CVA method suitable for multi-dimensional feature input needs to be proposed, which can maintain the differences between features, highlight the dominant features of each object, and maximise the retention of change information from multiple features, so that it can be expressed in multi-dimensional feature space, and all change vector directions can be fully reflected.

## 5.2 Theoretical Framework for Multi-feature Change Detection Based on SAW-CVA

The Self-Adaptive Weighted Change Vector Analysis (SAW-CVA) method developed in this book, which is adapted to multi-dimensional feature space, mainly includes two parts: one is the calculation of change intensity and change direction; the other is the introduction of self-adaptive weights (SAW) in the standardisation step for each feature. For different image objects, their corresponding weight factors are defined according to their different dominant features, maximising the retention of the change information of the features themselves. Without affecting the concentration of change vectors, the heterogeneity of change vectors is appropriately increased, enabling effective clustering and differentiation.

### *5.2.1 Change Intensity and Change Direction*

The intensity of change in SAW-CVA is also calculated through the modulus of the change vector $\Delta X$, but each component's $\Delta X$ is assigned an adaptive weight, reflecting the importance of change information within the multidimensional feature space. The specific calculation formula is as follows:

$$\rho = \sqrt{\sum_{n=1}^{N}(\omega_n \cdot (\Delta X_n)^2)}, \quad \left(0 < \omega_n < 1, \sum_{n=1}^{N} \omega_n = 1\right) \tag{5.4}$$

where, $\rho$ is the intensity of change for an object in the difference image; $\Delta X_n$ is the modulus of the $n$th change vector, i.e., the difference of the $n$th feature; $N$ is the total number of change vectors or features; $\omega_n$ is the adaptive weight of the $n$th feature, which controls the contribution of the feature to the change intensity. In the difference image, objects that have changed always have a higher change intensity than those that have not, so threshold segmentation can be used to distinguish between changed and unchanged image objects.

From the above analysis, it can be seen that the standard CVA, when facing two-dimensional change detection problems, uses formula (5.3) to calculate the direction of the change vector. However, this method cannot calculate all the change vector directions when facing change detection problems with multidimensional feature inputs. As can be seen from Fig. 5.2a, the change vector direction calculated by the standard CVA is based on the angle ($\theta_x$ or $\theta_y$) between the change vector and the coordinate axis. When the input features (where Vector $A$ and Vector $B$ represent input features $A$ and $B$) are not more than 2, the direction of their change vector indicates the content or type of change. But when the input features are more than 2, the direction of their change vector no longer contains any information about the type of change. Therefore, a fixed direction vector needs to be given as a reference vector, and the angle between the change vector and the reference vector is solved to measure the change direction.

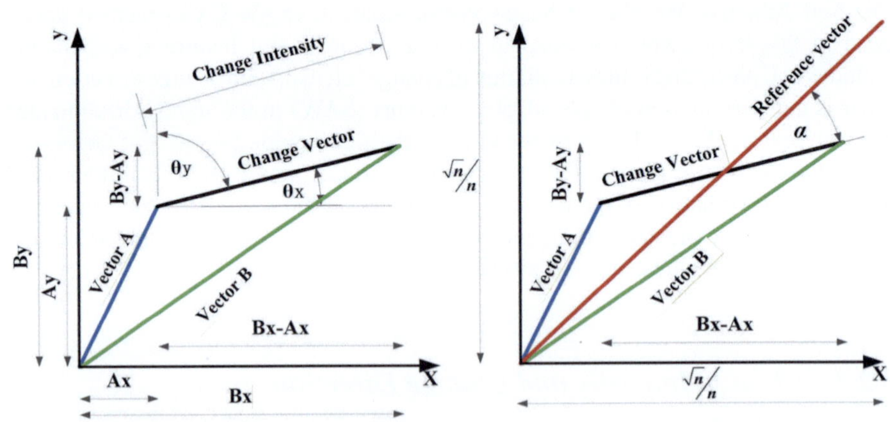


(a) Standard CVA (b) SAW-CVA

**Fig. 5.2** Two-dimensional schematic diagram of the direction of change vectors in standard CVA and SAW-CVA methods

The formula for the angle between two multidimensional vectors $t$ and $r$ in two-dimensional space is known to be:

$$\alpha = \arccos\left(\frac{\sum_{n=1}^{N}(t_n r_n)}{\sqrt{\sum_{n=1}^{N} t_n^2 \sum_{n=1}^{N} r_n^2}}\right), \quad (0 \le \alpha \le \pi) \tag{5.5}$$

where, $t_n$ and $r_n$ are the $n$th components of multidimensional vectors $t$ and $r$ respectively, vector $t$ is the change vector, $r$ is the reference vector; $\alpha$ is the angle between the change vector and the reference vector. But the reference vector is not easy to define and requires the assistance of prior knowledge. It can be approximately defined that the reference vector is composed of $N$ completely consistent unit vectors, so the starting point of the reference vector is the origin, and the end point coordinates are represented as $(\frac{\sqrt{n}}{n}, \frac{\sqrt{n}}{n})$, as shown in Fig. 5.2b. Therefore, formula (5.5) can be changed to

$$\alpha = \arccos\left(\frac{1}{\sqrt{N}}\left(\sum_{n=1}^{N}(\Delta X_n)/\sqrt{\sum_{n=1}^{N}(\Delta X_n)^2}\right)\right), \quad (0 \le \alpha \le \pi) \tag{5.6}$$

Similar to the calculation of change intensity, in order to highlight the differences of each feature and the dominant position of the main feature, an adaptive weight $\omega_n$ is assigned to each input feature vector, so formula (5.6) can be changed to

$$\alpha = \arccos\left(\sum_{n=1}^{N}(\omega_n \cdot \Delta X_n)/\sqrt{\sum_{n=1}^{N}(\omega_n \cdot (\Delta X_n)^2)}\right),$$
$$\left(0 \le \alpha \le \pi, 0 < \omega_n < 1, \sum_{n=1}^{N}\omega_n = 1\right) \tag{5.7}$$

where, $\alpha$ is the change direction of an object in the difference image; $\Delta X_n$ is the modulus of the $n$th change vector, i.e., the difference of the $n$th feature; $N$ is the total number of change vectors or features; $\omega_n$ is the adaptive weight of the $n$th feature, which controls the contribution of the feature to the change intensity. By classifying $\alpha$, different types of changed image objects can be identified.

### 5.2.2 *Adaptive Weight*

Generally speaking, the differences in each channel (band or feature component) of image objects at different times are determined by the differences in spectral and texture features of the objects, and the combination of spectral and texture differences is determined by adaptive weights (Li et al. 2014). This adaptive weight refers to the weight assigned to the features of image objects changing with the change of the dominant features of the image objects, that is, the adaptive weight is not constant, but is determined by the dominant features of the objects. In the process of object-oriented image analysis, many scholars have used similar adaptive weights, believing that such adaptive weights can well maintain the diversity and difference of image object features (Chen et al. 2002; Hu et al. 2005; Liang et al. 2014; Wang et al. 2010a). Wang once proposed an adaptive weight based on prior knowledge, which can better determine the dominant features of image objects, but due to the limitations of regional and empirical knowledge, the universality of the algorithm is relatively low (Wang et al. 2010a); Hu proposed an adaptive weight for estimating the contribution of texture, colour and brightness in the image segmentation process, and its estimation probability is obtained by comparing with a given threshold, but this given threshold is not objective and reliable.

For the determination of the dominant features of image objects, the approach adopted in this book is: the dispersion of each feature distribution of each image object in the multi-feature space can be obtained by measuring the internal standard deviation of the image objects, the internal standard deviation reveals the individual differences of these features, and by comparing the internal standard deviation, the dominant features of the image objects can be determined. For image objects whose dominant feature is spectral feature, their internal standard deviation is smaller; for image objects whose dominant feature is texture feature, their internal standard deviation is larger. Therefore, if the internal standard deviation of this image object is greater than a given threshold, it is considered that the dominant feature of this object is texture feature; if the internal standard deviation of this image object is less than a given threshold, it is considered that the dominant feature of this object is spectral feature. The setting of this given threshold depends on the whole image, the weight of the dominant feature should be greater than the weight of other features, it reflects the feature item with the most change information in the CVA input features.

Assuming $\Delta X_s$ and $\Delta X_t$ are the spectral difference component and texture difference component of an object in the difference image, respectively, then the difference $\Delta X$ of this image object can be expressed as

$$\Delta X = \sqrt{\omega_s(\Delta X_s)^2 + (1-\omega_s)(\Delta X_t)^2} \tag{5.8}$$

In the formula, $\omega_s$ is the adaptive weight of the spectral feature, then it is known that $(1-\omega_s)$ is the adaptive weight of the texture feature. Let $t_1$ and $t_2$ the internal standard deviations of the image objects at two time phases are $\text{std}_1$ and $\text{std}_2$, respectively, and the corresponding given thresholds are $k_1$ and $k_2$, then their spectral weight can be described as

$$\begin{gathered}\omega_s = \begin{cases} \max(\text{std}_1, \text{std}_2)/(\text{std}_1 + \text{std}_2), & \textit{if } \text{std}_1 < k_1 \text{ and } \text{std}_2 < k_2 \\ \min(\text{std}_1, \text{std}_2)/(\text{std}_1 + \text{std}_2), & \text{other} \end{cases} \\ k_i = \text{mean}_i + m_i \text{std}_i, \quad i = 1, 2 \end{gathered} \tag{5.9}$$

In the formula, mean $_i$ is $t_i$ the spectral mean of the image object; std $_i$ is $t_i$ the internal standard deviation of the image object; $m_i$ is an adaptive parameter, between the interval [− 2, 2].

The above analysis of the feature differences of each image object is for single-band images, and for multispectral images, the change information contained in each band should be fully utilized. Since the information entropy of each band can measure the amount of information contained in each band image, if the amount of information contained is larger, its entropy value will be larger, and the corresponding weight should be higher. In order to fully utilize the change information of each sub-feature in the spectral features and texture features, all input features in the spectral features and texture features are treated as a band, so the spectral difference component and texture difference component are

$$(\Delta X_s)^2 = \sum_{i=1}^{N_s} \omega_{si}(\Delta X_{si})^2, \quad (\Delta X_t)^2 = \sum_{i=1}^{N_t} \omega_{ti}(\Delta X_{ti})^2 \qquad (5.10)$$
$$\frac{\omega_{si}=|G_i|}{\sum_{i=1}^{N_s}|G_i|}, \quad \frac{\omega_{ti}=|G_i|}{\sum_{i=1}^{N_t}|G_i|}$$

In the formula, $N_s$ and $N_t$ represent the number of spectral and texture features respectively; $\omega_{si}$ and $\omega_{ti}$ represent the weights corresponding to each sub-feature in the spectral and texture features respectively; $G_i$ represents the information entropy of the $i$th sub-feature.

### 5.2.3 Polar Coordinate Solution of SAW-CVA

The Cartesian coordinate system is generally used to describe the distribution of CVA change vectors (Carvalho Júnior et al. 2011; Johnson et al. 1998; Nackaerts et al. 2005; Xian et al. 2010). If SAW-CVA is also presented in the form of a Cartesian coordinate system, it will be found that after introducing adaptive weights, both the magnitude and direction of the change vector have changed, thus identifying changes that the standard CVA could not identify before. Figure 5.3 shows the difference between the standard CVA and SAW-CVA change vectors in two-dimensional space. Figure 5.3a shows the distribution of each change vector in the standard CVA, with each axis representing the distribution of each feature in the difference image. For example, the vertical axis $\Delta F1$ represents the value of the difference image feature $F1$, which is the difference between the feature $F1$ of the two-time phase image objects, and the horizontal axis represents the difference of feature $F2$. Among the 8 change vectors, 3 are outside the threshold range (cv3, cv6, and cv8), so it can be considered that the image objects belonging to these 3 change vectors are the objects that have changed.

In Fig. 5.3b, since each feature (represented by each axis) is given an adaptive weight, the coordinate axis of SAW-CVA has changed, so $\Delta F1' = \omega_1 \times \Delta F1$, $\Delta F2' = \omega_2 \times \Delta F2$. At this time, the change vectors exceeding the threshold range become

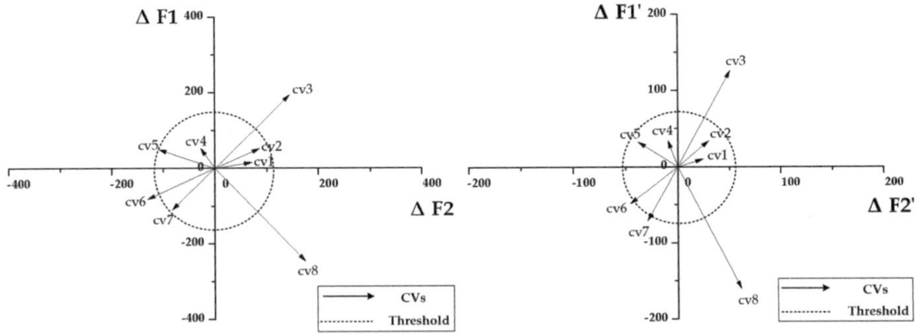


**Fig. 5.3** Comparison of Standard CVA and SAW-CVA in Cartesian coordinates

4 (cv3, cv6, cv7, and cv8). It can be seen that cv7 was not detected in the standard CVA.

The Cartesian coordinate system is only suitable for two-dimensional or three-dimensional CVA change detection problems. When facing multi-dimensional CVA, it will be difficult to display and describe. Based on the polar-representation proposed in the reference (Bovolo et al. 2007), the polar coordinate system is used to display the distribution of each change vector of multi-dimensional CVA. The display form of polar coordinates takes the change intensity $\rho$ as the radius and the change direction $\alpha$ as the angle, as shown in Fig. 5.4.

In Fig. 5.4, the entire coordinate system area is divided into 4 areas, where the change intensity threshold $T$ separates the changed area $SA_c$ (semi-annulus of changed area) and the unchanged area $SC_u$ (semi-circle of unchanged area), and $SA_c$ is divided into 3 areas $SA_{c1}$, $SA_{c2}$ and $SA_{c3}$ by the change direction thresholds $\alpha_1$ and $\alpha_2$. These areas can be described in mathematical expressions as follows

$$
\begin{aligned}
SC_u &= \{\rho, \alpha : 0 \le \rho \le T, 0 \le \alpha \le \pi\} \\
SA_{c1} &= \{\rho, \alpha : T \le \rho \le \rho 1_{\max}\{\}\} \\
SA_{c2} &= \{\rho, \alpha : T \le \rho \le \rho 1_{2\max}\{\}\} \\
SA_{c3} &= \{\rho, \alpha : T \le \rho \le \rho 2_{\max}\{\}\}
\end{aligned} \tag{5.11}
$$

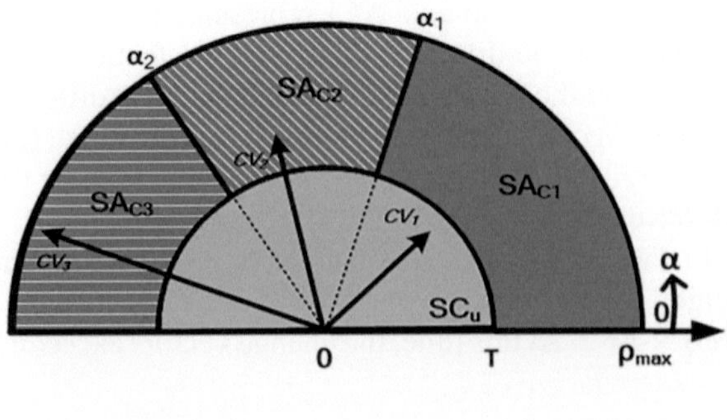


**Fig. 5.4** Polar coordinate solution of SAW-CVA

In the formula, $T$ is the change intensity threshold that distinguishes between changed and unchanged image objects in the difference image; $\alpha_1$ and $\alpha_2$ are the change direction thresholds that distinguish different types of changes in the changed image objects. Therefore, different areas in Fig. 5.4 contain different change information. For example, the change vector CV $_1$ implies that this image object is an unchanged object, while falling into $SA_{c2}$ of CV $_2$ implies that this image object is a changed object, and the type of change is c2. Therefore, it is known that the key to the change detection of multi-feature image objects based on SAW-CVA lies in $T$, as well as the determination of thresholds such as $\alpha_1$ and $\alpha_2$.

In summary, the theoretical framework of multi-feature OBCD based on SAW-CVA can fully exploit the advantages of multi-feature analysis and meet the needs for distinguishing and identifying multiple types of changes. The main features of this theoretical framework are:

(1) The difference in multi-features should be reflected in the detection of multi-feature changes in image objects, that is, different adaptive weights should be assigned to different feature input items;
(2) Automatic threshold segmentation of change intensity or other statistical measures expressing changes in differential image objects, thereby extracting changed objects;
(3) Unsupervised or supervised classification of statistical measures expressing the direction of change or other angles of change in changed objects, thereby meeting the needs for distinguishing and identifying multiple types of changes.

## 5.3 Multi-feature Change Detection of Image Objects Based on SAW-CVA

The multi-feature OBCD based on SAW-CVA mainly has two steps: ① Extraction of the change object area $SA_c$ based on the expectation maximisation threshold segmentation method; ② For the change area $SA_c$, through the distribution of direction variables, perform $K$ -means cluster analysis to identify each type of change area. This technical process is called the "two-step" detection mode of multi-feature change of high-resolution remote sensing image objects.

### *5.3.1 Extraction of Change Objects Based on Threshold Segmentation*

The extraction of change objects is based on a predefined threshold, distinguishing between changed and unchanged image objects, and the threshold is generally defined based on prior knowledge (Canty et al. 2006a; Chen et al. 2001). This way of defining the threshold has certain limitations, there will be a certain degree of subjective interference, and the acquisition of prior knowledge is also difficult. Common threshold segmentation methods include prior probability threshold method (Canty et al. 2006b; Doyle 1962), maximum inter-class variance method (Otsu 1975), histogram peak method (Xu et al. 2006), histogram curvature analysis method (Chen et al. 2015), entropy method (Kapur et al. 1985; Pal et al. 1989; Sahoo et al. 1997; Tescher 1988), iterative method (Reddi et al. 1984), Gaussian mixture model threshold method (Ridler et al. 1978) and double window variable step length method (Chen et al. 2001). In threshold segmentation algorithms, unsupervised methods can avoid human subjective interference and have a high degree of automation. This paper uses the Expectation Maximization (EM) algorithm in unsupervised methods to select thresholds, improving the automation and intelligence of the threshold selection process, making it more universal and general.

The ultimate standard of change detection is to maximise the change of the difference image, achieving separability from the unchanged part. Generally speaking, the changes and unchanged parts corresponding to the difference image obtained based on statistical analysis or linear transformation mostly follow a Gaussian distribution, so the Expectation Maximization algorithm in the Gaussian Mixture Model (GMM) (Landgrebe 1999) can be used for threshold segmentation. Unsupervised threshold segmentation can reduce subjective interference in the threshold selection process. The EM algorithm starts only from the data itself, without the need for prior knowledge, and is suitable for given observation data that follows a Gaussian distribution and is missing. The EM algorithm for parameter estimation was proposed by Dempster et al. (1977), it consists of the E-step and the M-step. The E-step is used to calculate the expectation of the log-likelihood function—Q function, and the M-step is used to select the parameters that maximise the expectation. After the M-step selects the parameters, the selected parameters are substituted into the E-step to calculate the expectation, and this is repeated until it converges to the optimal solution in the maximum likelihood sense. Its advantage is that it does not need to know the closed-form solution, and the calculation speed is relatively fast (Luo et al. 2002; Moon 1996) (Bruzzone et al. 2000b; Macqueen 1967). The use of EM for change threshold segmentation has been widely used in remote sensing image change detection research (Li et al. 2008; Bovolo et al. 2012; Jin et al. 2004; Liang et al. 2014).

Let the set of all objects in the image be $O = (O_1, O_2, \ldots, O_N)$, and $O_c$ and $O_u$ refer to the set of changed and unchanged objects, respectively. Let $P(O_c)$, $P(O_u)$, $p(\rho|O_c)$ and $p(\rho|O_u)$ represent the prior probabilities and conditional probability density functions of $O_c$ and $O_u$ respectively. Generally, the intensity of change corresponding to the changed object is greater than that of the unchanged object, so it is assumed that the intensity of object change approximately satisfies the Gaussian mixture distribution composed of two sub-Gaussians (Wang et al. 2010b), that is

$$p(\rho_i) = P(O_c)p(\rho_i|O_c) + P(O_u)p(\rho_i|O_u) \tag{5.12}$$

where, $I$ is the number of image objects, which is also the number of change vectors in SAW-CVA, then $i = (1, 2, \ldots, N)$, and $P(O_c) + P(O_u) = 1$. Assuming that the conditional probability density functions of the changed and unchanged classes of image objects both follow a Gaussian distribution, then

$$\frac{p(\rho_i|l) = \exp\left(-(\rho_i - \mu_l)^2/2\delta_l^2\right)}{\sqrt{2\pi\delta_l^2}} \tag{5.13}$$

where, $l \in (O_c, O_u)$, the parameters to be solved are the probability $p(l)$, mean $\mu_l$ and standard deviation $\delta_l$ corresponding to class $l$. Since the change category of each image object is unknown, this is a parameter solving problem with missing data, and the EM algorithm is suitable for solving such problems. The EM algorithm can obtain the maximum likelihood value in incomplete data in an iterative manner, and its iteration method is

$$\frac{p^{t+1}(l) = \sum_{i=1}^{I} p^t(l|\rho_i)}{I} \tag{5.14}$$

$$\frac{\mu_l^{t+1} = \sum_{i=1}^{I} p^t(l|\rho_i)\rho_i}{\sum_{i=1}^{I} p^t(l|\rho_i)} \tag{5.15}$$

$$\frac{\mu_l^{t+1} = \sum_{i=1}^{I} p^t(l|\rho_i)\rho_i}{\sum_{i=1}^{I} p^t(l|\rho_i)} \tag{5.16}$$

where, $t$ is the number of iterations; the initial value of EM is obtained through the $K$-means clustering algorithm. The process of solving the initial value parameters based on the $K$ -means clustering algorithm is as follows (Li et al. 2014): ① Use the $K$-means clustering algorithm to divide all image objects into $O_c$ and $O_u$ two categories, and obtain the initial values of the three parameters $p(l)$, $\mu_l$ and $\delta_l$; ② Perform iterations, use formulas (5.14)–(5.16) to update the values of these three parameters; ③ Repeat step ② until the difference between the parameters of two iterations is less than a given threshold. After obtaining the parameters corresponding to GMM, calculate the posterior probability of the image object belonging to $O_c$ and $O_u$:

$$\frac{p(l|\rho_i) = p(\rho_i|l)p(l)}{p(\rho_i)} \tag{5.17}$$

Based on the Bayes minimum error rate criterion, ignoring the constant $p(\rho_i)$ that is irrelevant to the category, each image object $O(i)$ can be categorised according to the following rules:

$$\begin{cases} p(\rho(i)|O_c)p(O_c) > p(\rho(i)|O_u)p(O_u),\ O_i \in O_c \\ p(\rho(i)|O_c)p(O_c) \le p(\rho(i)|O_u)p(O_u),\ O_i \in O_u \end{cases} \tag{5.18}$$

It is known that when $p(\rho(i)|O_c)p(O_c) = p(\rho(i)|O_u)p(O_u)$, the corresponding $\rho$ is the threshold $T$, which is the change intensity value corresponding to the intersection point of the probability curves of $O_c$ and $O_u$, as shown in Fig. 5.5.

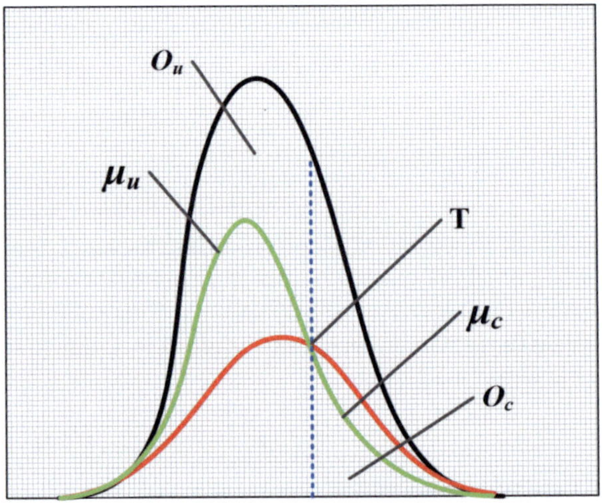


**Fig. 5.5** Schematic diagram of automatic threshold selection by EM algorithm

### *5.3.2 Identification of Change Types Based on Unsupervised Classification*

After extracting the change object area $SA_c$, it is necessary to consider how to distinguish the directions of various change vectors within the $SA_c$ area, that is, the identification of various change types of the change object. Assume the change object set $O_c = (O_{c1}, O_{c2}, \ldots, O_{ck})$, where $k$ is the number of change types, different change types correspond to different directions of change vectors, that is, the segmentation threshold of change direction is obtained through classification (there are $k$-1 segmentation thresholds). Due to the lack of prior knowledge, consider using unsupervised $K$-means clustering to obtain each segmentation threshold.

$K$ -means measures the Euclidean distance between the object and the cluster centre, continuously updates the cluster centre position in an iterative manner, and ends the iteration when certain conditions are met, thereby clustering the data (Macqueen 1967). Using $K$ -means, the change direction set $\alpha$ is divided into $k$ classes. Let $\alpha = (\alpha_1, \alpha_2, \ldots, \alpha_N)$, $N$ is the number of change vectors, the set of category $k$ is $C_k$, its cluster centre is $z_k$, then the objective function $J$ that satisfies the clustering condition is

$$J = \sum_{k=1}^{K} \sum_{i=1}^{n} \mu_k(\alpha_i)(d(\alpha_i, z_k))^2 \tag{5.19}$$

$$z_k = \frac{1}{N_k} \sum_{\forall \alpha_i \in C_k} \alpha_i \tag{5.20}$$

In the formula, $\mu_k(\alpha_i)$ is the membership function of $\alpha_i$ to $C_k$; $d(\alpha_i, z_k) = \|\alpha_i - z_k\|$ is the Euclidean distance from $\alpha_i$ to $z_k$. The objective function $J$ is established based on the minimum sum of squares error criterion, so $J$ measures the sum of squares error of $\alpha$ belonging to $C_k$, and the iteration ends continuously, the cluster centre changes continuously, and $J$ is as small as possible. At this time, find the maximum value within each category to determine the corresponding segmentation threshold. These segmentation thresholds ($\alpha_1$ and $\alpha_2$) can divide the SA $_c$ area, thus obtaining $SA_{c1}$, $SA_{c2}$ and $SA_{c3}$ in Fig. 5.4. These three regions, and each variation vector is also assigned to its corresponding region, then the type of change of the corresponding change object can be identified based on Eq. (5.11).

## 5.4 Experiment and Result Analysis

To analyse the effectiveness and reliability of the SAW-CVA method in multi-feature OBCD, two sets of experimental data were selected for verification, and compared with the standard CVA.

### *5.4.1 Experimental Area and Data*

The first set of experimental data uses FWV-2_2012 and FWV-2_2013, that is, the panchromatic and multispectral four-band fusion of WV-2_2012 and WV-2_2013 data. A portion of Area One (Beijing Olympic Core Area) was selected for the experiment, with 2500 × 1500 pixels and a resolution of 0.5 m. After preprocessing, image overlay was performed, and 512 image objects were obtained through MRS segmentation (Scale122, $w_{sh}$ 0.7, $w_{cp}$ 0.3).

Based on field investigation and visual judgement, appropriate reference data were selected. Three types of changes can be identified in the experimental area as shown in Fig. 5.6: $O_{c1}$ refers to objects where buildings or roads have changed, $O_{c2}$ refers to objects where artificial grasslands or other vegetation areas have changed, $O_{c3}$ refers to objects where changes are caused by sensor incidence angle offset. It should be pointed out that, $O_{c3}$ is not a change in the true sense, but a difference between two images caused by different sensor shooting angles, the most direct display is the change of shadows of high-rise buildings. The reference data contains 347 reference objects belonging to $O_u$, 165 reference objects belonging to $O_{c1}$, 53 reference objects belonging to $O_{c2}$, and 112 reference objects belonging to $O_{c3}$. The experimental area is located in the city centre, with complex man-made features, fine change areas, and some pseudo-changes or fine changes are not easy to identify. The purpose of the experiment is to compare and verify the differences between SAW-CVA and standard CVA in fine change and multi-change type recognition.

The second experiment focuses on the changes in agricultural land. The experimental area is still Area Two (Beijing Changping Agricultural Demonstration Garden), which is an area with agricultural land as the main land use type. The

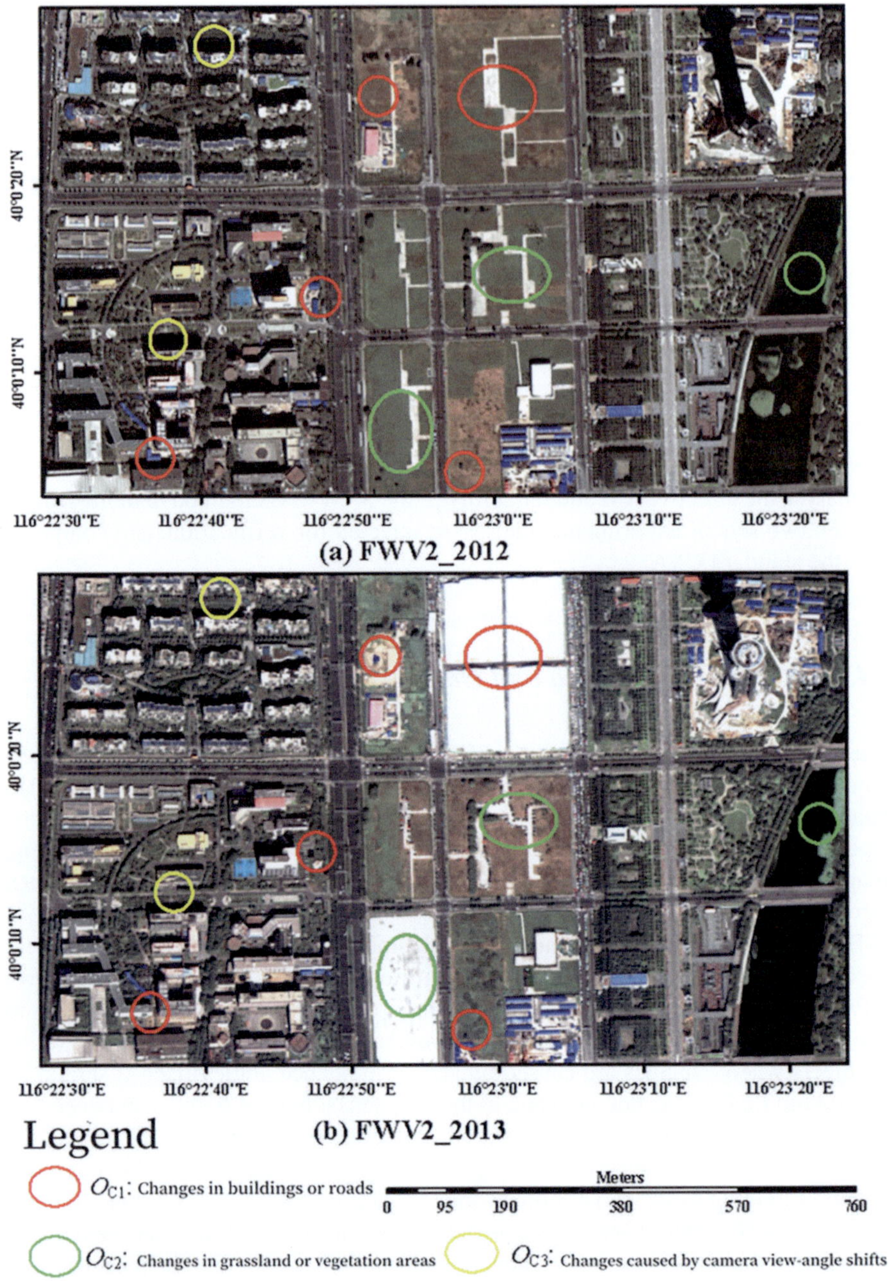


**Fig. 5.6** Area One WV-2 fusion true colour composite image data (colour map attached later)

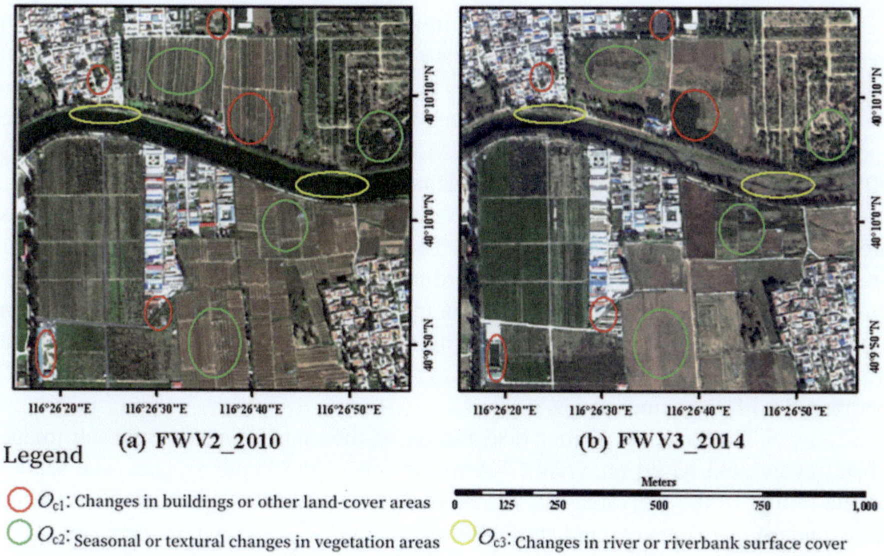


**Fig. 5.7** Area Two WV-2/3 fusion true colour composite image data (colour map attached later)

data selected are FWV-2_2010 and FWV-3_2014, four-band fusion images of 500 × 500 pixels. Based on MRS image segmentation (Scale196, $w_{sh}$ 0.4, $w_{cp}$ 0.4), 432 objects were obtained (Fig. 5.7).

The main types of changes that can be identified in Experimental Area Two are: $O_{c1}$ refers to changes in coverage of buildings or other areas, such as changes where new buildings have been built in farmland areas, or changes where buildings have become forest or bare land; $O_{c2}$ refers to seasonal or texture pattern changes in farmland or other vegetation areas, such as changes in crop seasons or farming methods, texture pattern changes caused by field ridge replacement, and changes in returning farmland to forest; $O_{c3}$ refers to geometric changes in rivers or riverbanks, such as changes in coverage caused by changes in river water levels, and changes in coverage of aquatic plants or other aquatic plants caused by river eutrophication. Among these three changes, $O_{c1}$ is relatively easy to identify, while the other two types of changes are more difficult to distinguish. The reference data contains 347 reference objects belonging to $O_u$, 165 reference objects belonging to $O_{c1}$, 53 reference objects belonging to $O_{c2}$, and 112 reference objects belonging to $O_{c3}$.

### 5.4.2 *Threshold Segmentation of Change Intensity*

The generation of change vectors is based on the combined effect of multiple features of the image object. As known from Chap. 4, there are various types of image object features. Feature extraction can establish an image object feature library, and the GPSO feature optimisation selection algorithm can automatically select a multi-feature combination suitable for OBCD.

For the first set of experiments, the image object feature library is as shown in Table 4.1, with the number of candidate features $M = 7$, the initial population of GPSO is 80, the learning factor $c_1 = 2.6$ and $c_2 = 1.6$, the inertia weight $\omega = 0.9$, the maximum number of iterations $iter_{max} = 100$. The input features are spectral mean, standard deviation, GLCM-Correlation (0°), GLCM-Contrast (0°, 90°), GLCM-Entropy (0°) and GLCM-Mean (0°).

According to the introduction of SAW-CVA in Sect. 5.3, the optimised features are used as input features into SAW-CVA. The two sets of experiments calculate their respective feature adaptive weights according to formulas (5.8) to (5.10), calculate the change intensity according to formula (5.4), and calculate the change direction according to formula (5.7), thus obtaining the change vector distribution of each image object. This is displayed in the form of scatter plots in polar coordinates and compared with the standard CVA.

Figure 5.8a shows the scatter distribution of the change vectors of each image object calculated based on SAW-CVA in the polar coordinate system. The scatter points falling in the $SA_c$ area (dark grey area) are considered as image objects that have undergone changes, and those falling in the $SC_u$ area (light grey area) are considered as image objects that have not undergone changes. These two areas are divided by the threshold $T$. According to Sect. 5.4.1, the threshold $T$ is obtained by the EM algorithm.

Figure 5.9a shows a comparison of the estimated change intensity probability density curve (blue curve) based on the EM algorithm and the actual change intensity histogram (red bar) distribution. It can be seen that the blue curve and the red bar have a high degree of fit, indicating that the parameters obtained based on the EM

**Fig. 5.8** Threshold segmentation of change intensity based on SAW-CVA and standard CVA (Experiment One)

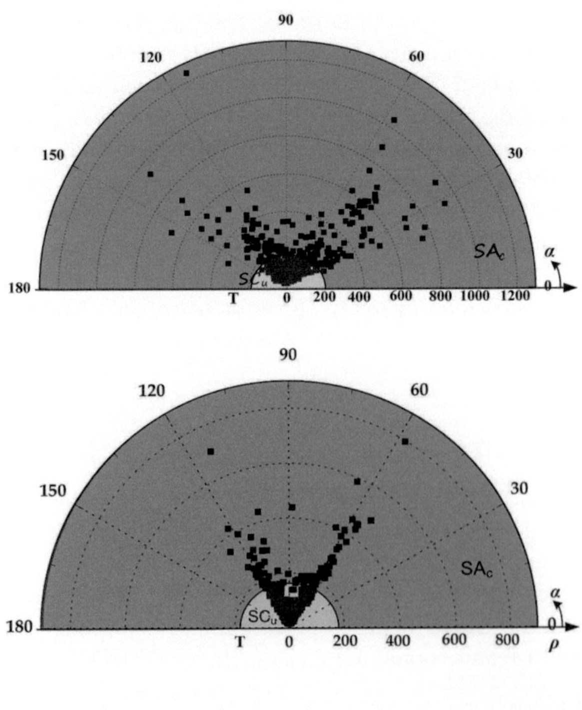

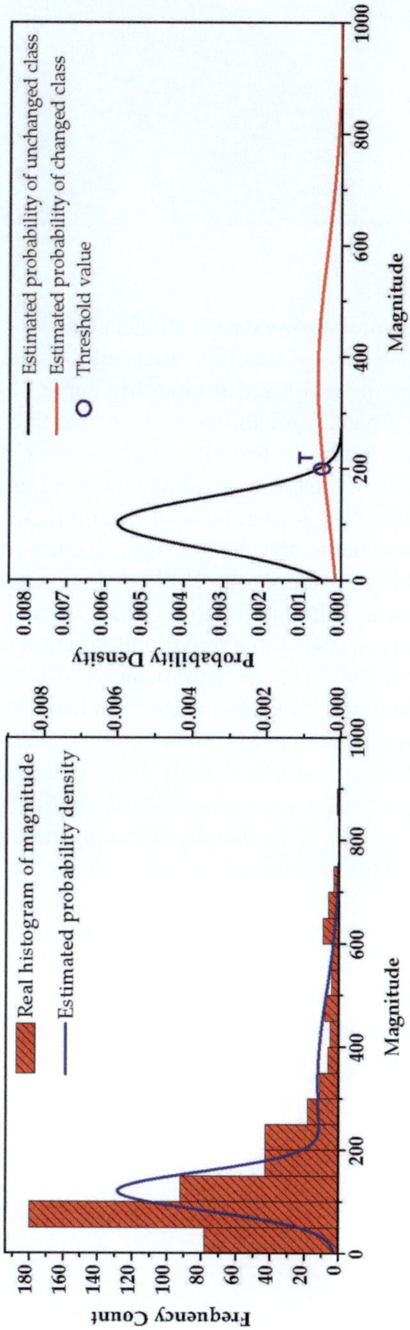


**Fig. 5.9** Estimated change intensity probability density curve based on EM algorithm and probability curves of $O_c$ and $O_u$ (Experiment One; colour figure attached)

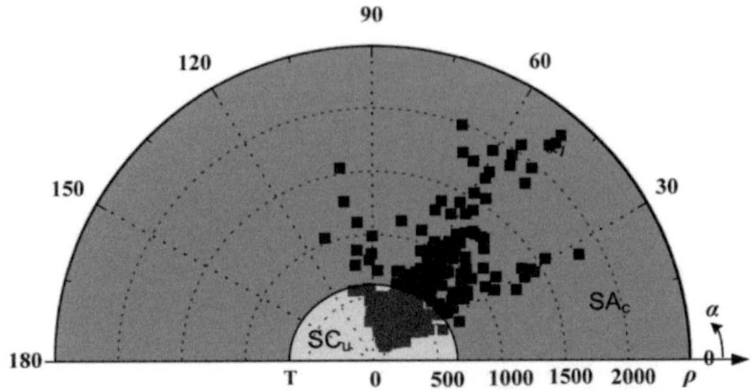


**Fig. 5.10** Change intensity threshold segmentation based on SAW-CVA and standard CVA (experiment two)

algorithm are reliable, and the probability curves of each category constructed from this are reliable. On this basis, the probability areas of $O_c$ and $O_u$ are shown in Fig. 5.9b. The red curve is the estimated probability curve of the change class, and the black curve is the estimated probability curve of the unchanged class. The two curves intersect at $T$. According to formula (5.18) in Sect. 5.4.1, at this time $P(O_c)p(\rho_i|O_c) = P(O_u)p(\rho_i|O_u)$, the corresponding change intensity value $T$ is the threshold to distinguish $O_c$ and $O_u$. As can be seen from Fig. 5.9, $T = 200$.

For the second set of experiments, the image object feature library is as shown in Area Two of Table 4.1, with the number of candidate features $M = 7$. The initial parameters of GPSO are set to an initial population of 60, learning factors $c_1 = 2.8$ and $c_2 = 1.3$, inertia weight $\omega = 0.9$, with a maximum iteration count of $iter_{max} = 80$. The optimised features obtained are spectral mean, NDVI, GLCM-Correlation (0°), GLCM-Contrast (0°, 90°) and GLCM-Ang.2nd moment (90°). Similar to the first set of experiments, the optimised features are input into SAW-CVA. Both sets of experiments calculate the adaptive weights of their respective features according to Eqs. (5.8)–(5.10), calculate the change intensity according to Eq. (5.4), and calculate the change direction according to Eq. (5.7), thereby obtaining the change vector distribution of each image object. This is displayed in a scatter plot in polar coordinates as shown in Fig. 5.10.

As with the first set of experiments, as can be seen from Fig. 5.10, the $SA_c$ region and the $SC_u$ region are divided by the threshold $T$. Figure 5.11a reveals that the change intensity probability density curve estimated based on the EM algorithm fits very well with the actual change intensity histogram distribution, so the threshold $T = 600$ can be estimated according to Fig. 5.11b.

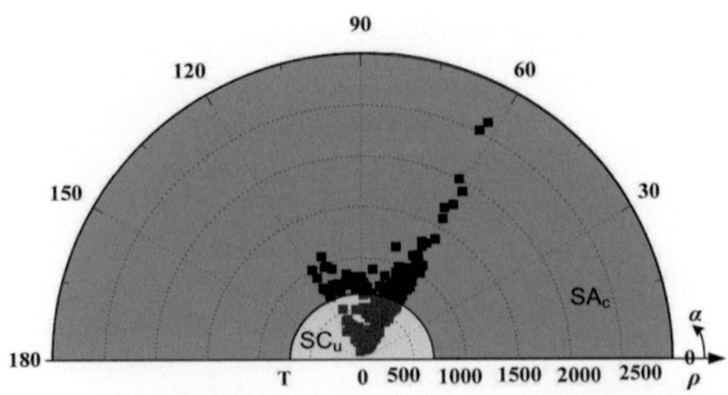


**Fig. 5.11** Change intensity probability density curve estimated based on the EM algorithm and the probability curves of $O_c$ and $O_u$ (experiment two; colour figure follows)

### 5.4.3 Distinguishing Change Directions

The identification of change types mainly relies on the $K$ -means clustering analysis of change directions, obtaining the corresponding segmentation thresholds for type division. For the first set of experiments, as shown in Fig. 5.12a, the scatter points based on SAW-CVA can be clustered into four categories (curve range), which are divided according to three direction thresholds ($\alpha_1$, $\alpha_2$, $\alpha_3$). Each cluster category corresponds to a type of change, according to Eqs. (5.19) to (5.20), where $K = 4$, the segmentation thresholds are $\alpha_1 = 42°$, $\alpha_2 = 78°$ and $\alpha_3 = 113°$. Combined with the change intensity segmentation threshold $T = 200$ calculated in Sect. 5.5.2, the four regions shown in Fig. 5.12a can be expressed as:

$$\begin{aligned} \mathrm{SC}_u &= \{\rho, \alpha : 0 \le \rho \le 200, 0° \le \alpha \le 180°\}\mathrm{SA}_{c1} \\ &= \{\rho, \alpha : 200 \le \rho \le 1250, 0° \le \alpha \le 42°\}\mathrm{SA}_{c2} \\ &= \{\rho, \alpha : 200 \le \rho \le 1250, 42° \le \alpha \le 78°\}\mathrm{SA}_{c3} \\ &= \{\rho, \alpha : 200 \le \rho \le 1250, 78° \le \alpha \le 113°\} \\ \mathrm{SA}_{c4} &= \{\rho, \alpha : 200 \le \rho \le 1250, 113° \le \alpha \le 180°\} \end{aligned} \tag{5.21}$$

In the formula, $\mathrm{SC_u}$, $\mathrm{SA_{c1}}$, $\mathrm{SA_{c2}}$, $\mathrm{SA_{c3}}$ and $\mathrm{SA_{c4}}$ correspond one-to-one with $O_\mathrm{u}$, $O_\mathrm{c1}$, $O_\mathrm{c2}$, $O_\mathrm{c3}$ and $O_\mathrm{c4}$ respectively. Compared with SAW-CVA, the scatter distribution of standard CVA is relatively concentrated (Fig. 5.12b), can only be clustered into two categories, and there is a significant difference from SAW-CVA. Moreover, the change intensity threshold is 180, and there is only one change direction threshold of 83°.

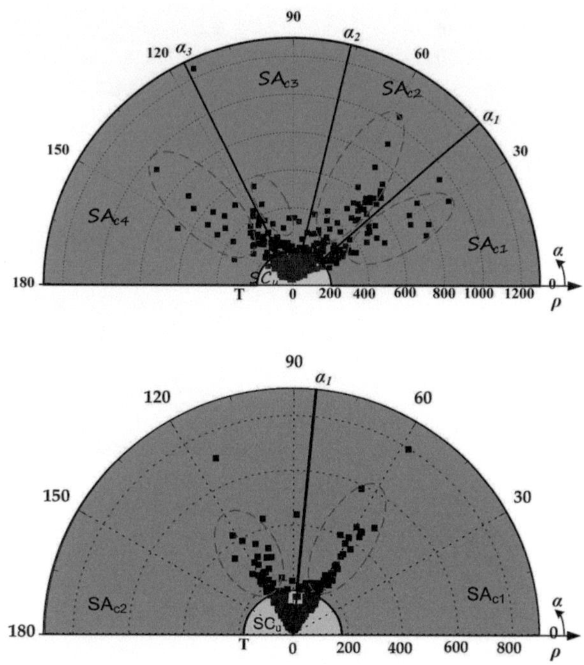


**Fig. 5.12** Change direction classification based on SAW-CVA and standard CVA (Experiment One)

For the second set of experiments, as shown in Fig. 5.13a, the scatter points based on SAW-CVA can be clustered into four categories (curve range), which are divided according to two direction thresholds ($\alpha_1$, $\alpha_2$). Each cluster category corresponds to a type of change, according to Eqs. (5.19) to (5.20), where $K = 3$, the segmentation thresholds are $\alpha_1 = 42°$ and $\alpha_2 = 80°$. Combining with the change intensity segmentation threshold $T = 600$ calculated in Sect. 55.5.2, the four regions shown in Fig. 5.13a can be expressed:

$$\begin{aligned} \mathrm{SC}_u &= \{\rho, \alpha : 0 \le \rho \le 600, 0° \le \alpha \le 180°\}\mathrm{SA}_{c1} \\ &= \{\rho, \alpha : 600 \le \rho \le 2500, 0° \le \alpha \le 42°\}\mathrm{SA}_{c2} \\ &= \{\rho, \alpha : 600 \le \rho \le 2500, 42° \le \alpha \le 80°\} \\ \mathrm{SA}_{c3} &= \{\rho, \alpha : 600 \le \rho \le 2500, 80° \le \alpha \le 180°\} \end{aligned} \tag{5.22}$$

where, $\mathrm{SC_u}$, $\mathrm{SA_{c1}}$, $\mathrm{SA_{c2}}$ and $\mathrm{SA_{c3}}$ correspond to $O_\mathrm{u}$, $O_\mathrm{c1}$, $O_\mathrm{c2}$ and $O_\mathrm{c3}$ respectively. Compared with SAW-CVA, the scatter distribution of standard CVA is relatively concentrated (Fig. 5.13b), can only be grouped into two types, which is significantly different from SAW-CVA. The segmentation threshold for change intensity is 680, and there is only one segmentation threshold for change direction, 84°.

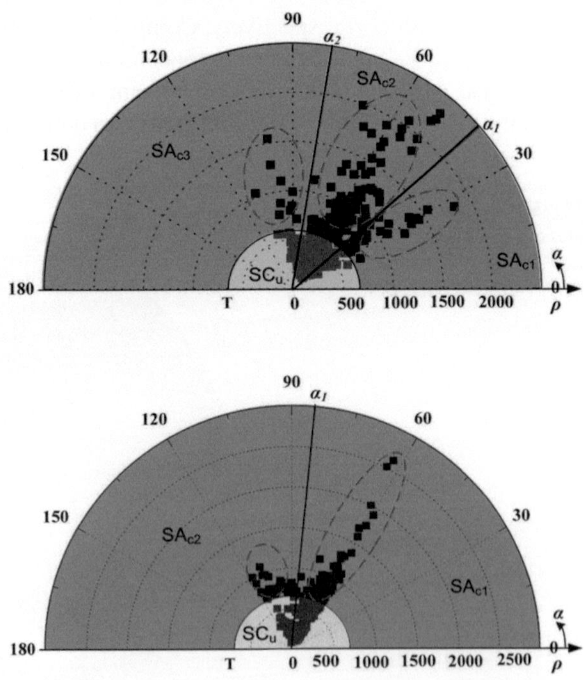


**Fig. 5.13** Direction classification based on SAW-CVA and standard CVA (Experiment Two)

### 5.4.4 Change Detection Results and Accuracy Evaluation

Using the segmentation thresholds (including change intensity and direction) obtained in Sects. 5.5.2 and 5.5.3, the difference image objects are identified to obtain multi-feature OBCD results, and accuracy evaluation is carried out in combination with reference data.

1. Results and accuracy evaluation of the first set of multi-feature OBCD experiments

Figure 5.14 shows the results of the first set of experiments. The image objects with fill colours are the change objects, and those without colours are the unchanged objects. SAW-CVA can identify more types of changes than standard CVA. It is known that there are 355 unchanged objects and 157 changed objects in the results of SAW-CVA, of which 49 objects belong to $O_{c1}$, 52 objects belong to $O_{c2}$, 44 objects belong to $O_{c3}$, and 12 objects belong to $O_{c4}$. Comparing with the reference data, the accuracy of the OBCD results of SAW-CVA is evaluated, and the error confusion matrix is calculated to obtain the accuracy evaluation results (Tables 5.1 and 5.2).

**Table 5.1** Accuracy evaluation results of multi-feature OBCD based on SAW-CVA (Experiment One)

| SAW-CVA | | Reference data | | | | | Misjudgement rate (%) |
|---|---|---|---|---|---|---|---|
| | | $O_u$ | $O_{c1}$ | $O_{c2}$ | $O_{c3}$ | $O_{c4}$ | |
| Result data | $O_u$ | 346 | 4 | 3 | 2 | 0 | 2.54 |
| | $O_{c1}$ | 2 | 38 | 6 | 3 | 0 | 22.45 |
| SAW-CVA | | Reference data | | | | | Misjudgement rate (%) |
| | | $O_u$ | $O_{c1}$ | $O_{c2}$ | $O_{c3}$ | $O_{c4}$ | |
| Result data | $O_{c2}$ | 6 | 5 | 36 | 5 | 0 | 30.77 |
| | $O_{c3}$ | 8 | 2 | 2 | 32 | 0 | 27.27 |
| | $O_{c4}$ | 10 | 1 | 0 | 1 | 0 | 0.00 |
| Miss rate (%) | | 6.99 | 24.00 | 23.40 | 25.58 | 0.00 | |
| Overall accuracy | | 88.28% | | Kappa | 0.75 | | |

**Table 5.2** Accuracy evaluation results of multi-feature OBCD based on standard CVA (Experiment One)

| CVA | | Reference Data | | | | Misjudgement rate (%) |
|---|---|---|---|---|---|---|
| | | $O_u$ | $O_{c1}$ | $O_{c2}$ | $O_{c3}$ | |
| Result data | $O_u$ | 356 | 6 | 4 | 9 | 4.81 |
| | $O_{c1}$ | 3 | 38 | 2 | 24 | 43.28 |
| | $O_{c2}$ | 13 | 6 | 41 | 10 | 41.43 |
| | $O_{c3}$ | 0 | 0 | 0 | 0 | 0.00 |
| Miss rate (%) | | 4.30 | 24.00 | 12.77 | 100.00 | |
| Overall accuracy | | 84.96% | | Kappa | 0.66 | |

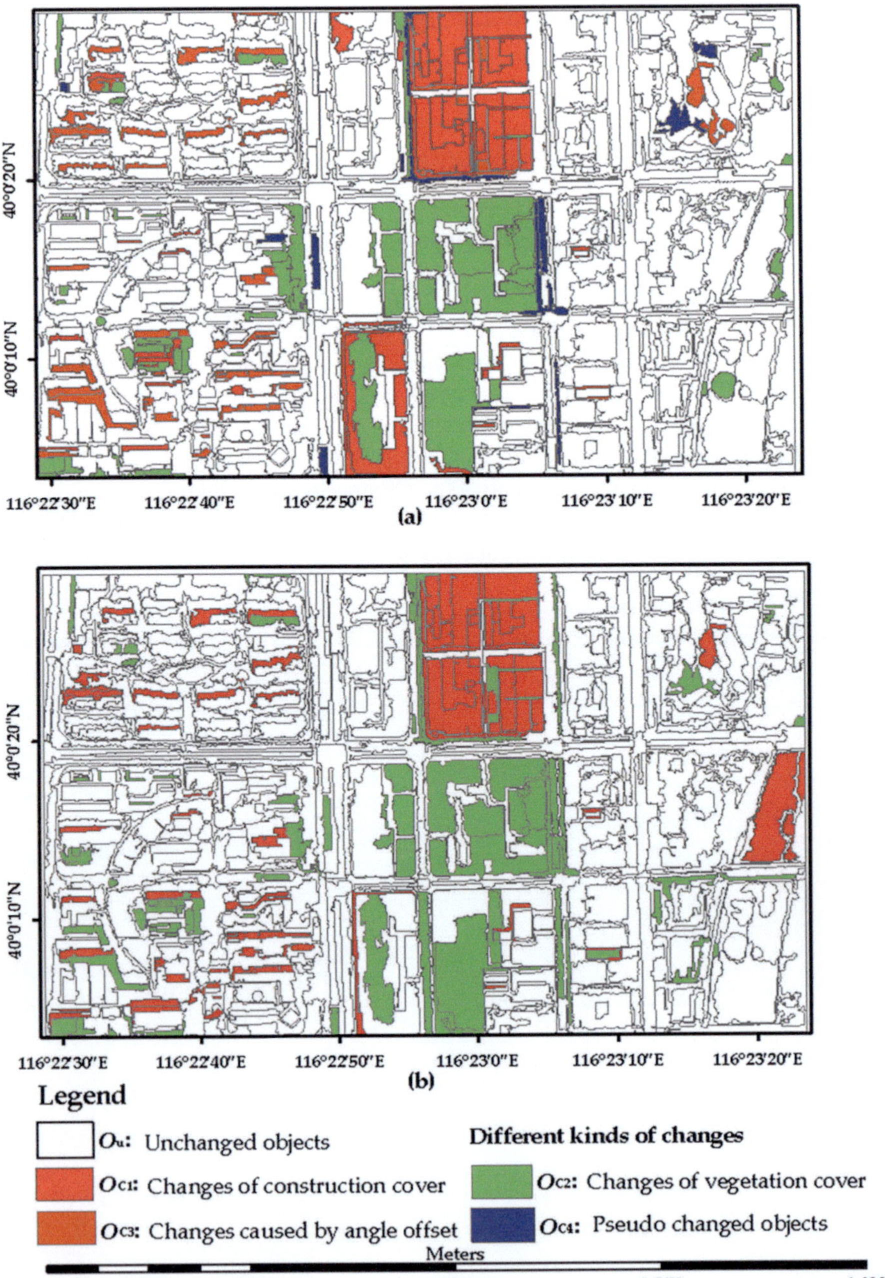


**Fig. 5.14** OBCD results based on SAW-CVA and standard CVA (Experiment One; colour map attached later)

2. Results and Accuracy Evaluation of Multi-Feature OBCD for the Second Set of Experiments

Figure 5.15 shows the multi-feature OBCD results of the second set of experiments based on SAW-CVA and standard CVA. Similar to the first set of experiments, there are three types of changes in the results of SAW-CVA, while there are only two types of changes in the results of standard CVA. Statistics show that there are 182 unchanged objects and 250 changed objects in the results of SAW-CVA, of which 128 objects belong to $O_{c1}$, 90 objects belong to $O_{c2}$, and 32 objects belong to $O_{c3}$.

Tables 5.3 and 5.4 show the results of the second set of experiments. Similar to the first set of experiments, it can be seen that the overall accuracy and Kappa coefficient of SAW-CVA (86.03%, 0.80) are higher than those of standard CVA (75.86%, 0.63). The standard CVA cannot identify changes caused by seasonal changes or texture patterns, and incorrectly classifies these changes into other types of changes, so it cannot accurately distinguish between fine changes and pseudo-changes. Table 5.3 also shows that in the results of SAW-CVA, $O_{c1}$ has a higher miss rate (25%), $O_{c2}$ has a higher misjudgment rate (31.63%). It can be seen that some objects belonging to $O_{c1}$ are misclassified as $O_{c2}$. This indicates that although SAW-CVA can identify two different types of changes, $O_{c1}$ and $O_{c2}$, these two types of image objects have similar features and are easily misclassified. This can also be seen from Fig. 5.13a, where the scatter of these two types is very close.

The experiment proves that compared with standard CVA, SAW-CVA can identify more types of changes, especially some fine changes, proving that the segmentation

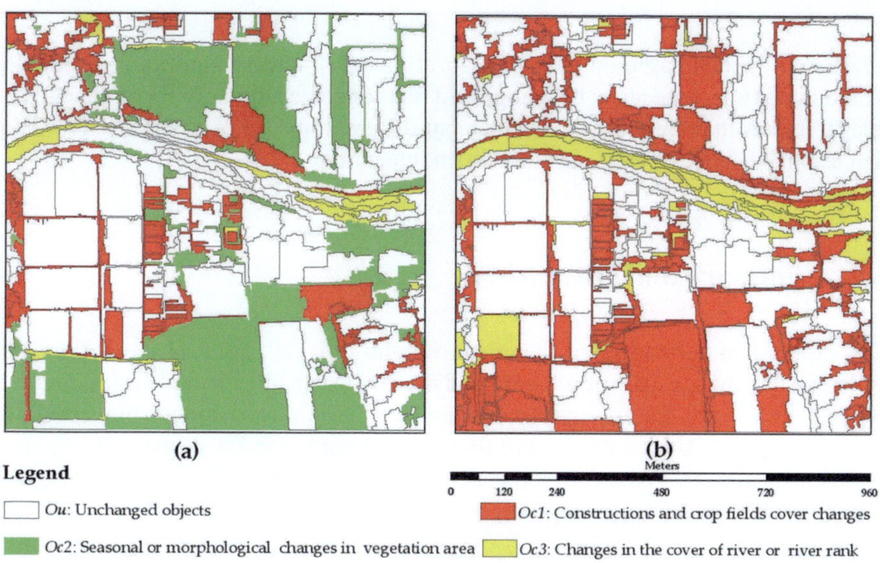


**Fig. 5.15** OBCD results based on SAW-CVA and standard CVA (experiment two; colour figure attached)

**Table 5.3** Accuracy evaluation results of multi-feature OBCD based on SAW-CVA (experiment two)

| SAW-CVA | | Reference data | | | | Misjudgment rate (%) |
|---|---|---|---|---|---|---|
| | | $O_u$ | $O_{c1}$ | $O_{c2}$ | $O_{c3}$ | |
| Result data | $O_u$ | 168 | 4 | 4 | 2 | 5.56 |
| | $O_{c1}$ | 4 | 96 | 2 | 2 | 8.24 |
| | $O_{c2}$ | 10 | 26 | 83 | 2 | 31.63 |
| | $O_{c3}$ | 0 | 2 | 1 | 25 | 13.04 |
| Miss rate (%) | | 7.48 | 25.00 | 8.22 | 23.08 | |
| Overall accuracy | | 86.03% | | Kappa | 0.80 | |

**Table 5.4** Accuracy evaluation results of multi-feature OBCD based on standard CVA (experiment two)

| Standard CVA | | Reference Data | | | | Misjudgement Rate (%) |
|---|---|---|---|---|---|---|
| | | $O_u$ | $O_{c1}$ | $O_{c2}$ | $O_{c3}$ | |
| Result data | $O_u$ | 160 | 2 | 31 | 0 | 17.20 |
| | $O_{c1}$ | 12 | 123 | 47 | 1 | 32.89 |
| | $O_{c2}$ | 0 | 0 | 0 | 0 | 0.00 |
| | $O_{c3}$ | 9 | 2 | 12 | 31 | 43.18 |
| Miss rate (%) | | 11.56 | 3.85 | 100.00 | 3.85 | |
| Overall accuracy | | 75.86% | | Kappa | 0.63 | |

threshold obtained based on the EM algorithm and $K$ -means algorithm is reliable. At the same time, it needs to be recognized that for some similar types of changes or image objects that are close in distance but different in type of change, SAW-CVA will also have a higher miss rate and misjudgment rate.

## 5.5 Summary and Discussion

The experiment demonstrated the advanced and reliable nature of the SAW-CVA multi-feature change detection theory and the "two-step" approach, highlighting the advantages of multi-feature analysis and meeting the need for distinguishing multiple types of changes. The following will focus on the compatibility of the multi-feature change detection theoretical framework with other change detection methods, and the universality of the "two-step" approach.

### 5.5.1 Comparison of Different Multi-feature OBCD Methods

This section discusses the differences in its approach to various change detection methods, implementing change object extraction and change type distinction recognition through the "two-step" approach. The selected change detection methods are the object-based hybrid multivariate alteration detection (OB-HMAD) and the object-based multi-feature fuzzy fusion (OB-MFF) algorithms.

1. Change Detection Based on OB-HMAD

Based on the multivariate alteration detection (MAD) algorithm of canonical correlation analysis, using object-oriented technology, the canonical correlation (MAD) transformation is carried out based on image objects. To make the change information on the MAD variable difference image more prominent and improve the separability of categories, a minimum noise fraction (MNF) transformation is performed. Finally, on the first component of MNF, binary threshold segmentation is used to extract change objects. This method, which integrates object-oriented thinking and MNF transformation on the basis of the MAD algorithm, is called the OB-HMAD algorithm.

Suppose two temporal remote sensing images $T_1$ and $T_2$ are obtained after object-oriented multi-scale segmentation, and the feature mean values of each channel object corresponding to random variables $X$ and $Y$ are extracted, with an expected mean of 0, and are linearly transformed into

$$U = a^T X = a_1 X_1 + \cdots a_p X_p \quad (5.23)$$

$$V = b^T Y = b_1 Y_1 + \cdots b_p Y_p \quad (5.24)$$

Different linear combinations $U$ and $V$ are composed of different coefficient vectors $a$ and $b$, and the MAD transformation about $X$ and $Y$ is to find vectors $a$ and $b$ from this set of linear transformations that maximise $\mathrm{Var}(U - V)$, with the constraint condition $\mathrm{Var}(U) = \mathrm{Var}(V) = 1$. The corresponding $U$ and $V$ are called the canonical variables of $X$ and $Y$, and the correlation coefficients $a$ and $b$ are called the canonical change coefficients. It is known that under the constraint condition:

$$\mathrm{Var}(U - V) = \mathrm{Var}\left(a^T X\right) + \mathrm{Var}\left(b^T Y\right) - 2\mathrm{Cov}\left(a^T X, b^T Y\right) = 2\left(1 - \mathrm{Cov}\left(a^T X, b^T Y\right)\right) \quad (5.25)$$

The difference image obtained through the MAD transformation concentrates most of the change information. The canonical variables obtained from the canonical transformation can be arranged in ascending order according to the canonical correlation coefficient, from which the MAD variables can be obtained:

$$\mathrm{MAD}_i = U_{p+1-i} - V_{p+1-i} = a^T_{p+1-i} X - b^T_{p+1-i} Y \quad (5.26)$$

The MAD variables are subjected to a minimum noise transformation to effectively separate the change information from the noise in the MAD variables. By using the maximum signal-to-noise ratio to effectively concentrate the change information, the difference information in the MAD variables is separated (Green et al. 1994). After the MNF transformation, the MNF variables replace the MAD variables, thus constructing a difference image with more concentrated change information. Generally speaking, the first MNF variable has the largest signal-to-noise ratio and contains the most change information.

After the OB-MAD/MNF transformation, the multi-temporal images are transformed into difference images where the change information is concentrated. On this image layer, the object spectral features present a normal distribution, the closer the object area is to the middle peak, the lower the possibility of change, and the farther the area is from the peak ends, the higher the possibility of change. The threshold method can be used to distinguish between changed and unchanged objects. The OB-HMAD method can accurately extract change objects, but cannot determine the type of object change.

2. Change Detection Based on OB-MFF

The representation of multi-feature variation information is often reflected through the differences in features, with different feature inputs yielding different OBCD results. This inconsistency is uncertain information that cannot be directly measured by a single value, but can be explained through fuzzy set theory. Fuzzy set theory describes the specific elements in an event in the form of membership through the analysis of uncertain information of the event (Ghosh et al. 2009). This book refers to the fuzzy fusion strategy in (Carvalho et al. 2013), applying fuzzy set theory to the fusion of different types of multi-feature variation information of image objects. By constructing fuzzy membership functions of different feature difference image objects for different categories (changed or unchanged), the category membership of different features of the object is calculated. The final membership of each object is obtained through fuzzy weighted fusion. According to the principle of maximising membership, the fuzzy quantity is turned into a certain quantity, and its category is determined. This change detection method is called the change detection method based on multi-feature fuzzy fusion of image objects (OB-MFF), and the basic process of OB-MFF is as follows (Carvalho et al. 2013). Firstly, image objects are obtained through image segmentation, and the difference image X of two time phases is calculated. This image contains N-dimensional feature bands, and the number of image objects is M. The segmentation threshold Tk of each feature difference image is obtained through the EM or other threshold segmentation algorithms, where k is the number of features. Secondly, based on the threshold $T_k$, the membership $H_c(X_k(i))$ of the ith object $X_k(i)$ in the kth feature difference image $X_k$ relative to the change class Oc, and the membership $H_u(X_k(i))$ relative to the unchanged class Ou, are calculated. The S-shaped function is used to calculate the value of the membership, i.e.,

$$H_c(X_k(i)) = \begin{cases} 0, & X_k(i) \le a_k \\ 2\left\{\frac{X_k(i)-a_k}{c_k-a_k}\right\}^2, & a_k \le X_k(i) \le b_k \\ 1-2\left\{\frac{c_k-X_k(i)}{c_k-a_k}\right\}^2, & b_k \le X_k(i) \le c_k \\ 1, & X_k(i) \ge c_k \end{cases} \tag{5.27}$$

$$H_u(X_k(i)) = 1 - H_c(X_k(i)) \tag{5.28}$$

where, $a_k = 0.8T_k, c_k = T_k, b_k = (a_k + c_k)/2$, the closer $H_c(X_k(i))$ is to 1, the greater the possibility that $X_k(i)$ belongs to the change class, and vice versa, the greater the possibility that it belongs to the unchanged class. Thirdly, based on the category membership of different features, the final membership of the object can be obtained through fuzzy weighted fusion:

$$\begin{cases} H_c(i) = \sum_{k=1}^{N} w_k \times H_c(X_k(i)) \\ H_u(i) = \sum_{k=1}^{N} w_k \times H_u(X_k(i)) \end{cases} \tag{5.29}$$

In the formula, $H_c(i)$ is the membership of the ith object belonging to the change class; $w_k$ is the kth feature. In most literature, each feature is assigned the same weight (Carvalho et al. 2013; Castilla and Hay 2007), i.e., $w_k = 1/N$. However, according to the first principle of the multi-feature OBCD theoretical framework in Sect. 5.1, the differences between features should be reflected. The weight selection here uses the adaptive weights from Sect. 5.2.2. Finally, the changed objects on the difference image are extracted according to the maximum membership discrimination rule: $O_c$.
3. Comparison of different OBCD methods Under the guidance of the multi-feature OBCD theoretical framework and the "two-step" model, this book improves the OB-HMAD and OB-MFF methods to take advantage of multi-feature analysis and distinguish different types of changes in changed objects. The input features should reflect the differences as much as possible. For the OB-HMAD method, adaptive weights need to be assigned to each variable before the MAD transformation, so that the input features/bands maintain different weights:

$$U = a^T wX = a_1 w_1 X_1 + \cdots a_p w_p X_p \tag{5.31}$$

$$V = b^T wY = b_1 w_1 Y_1 + \cdots b_p w_p Y_p \tag{5.32}$$

w here $w$ is the adaptive weight assigned to each feature, the specific calculation method refers to Sect. 5.2.2. For the OB-MFF algorithm, similar weights are like $w_k$ in Eq. (5.29). The extraction of changed objects mainly relies on automated threshold segmentation methods. For SAW-CVA and standard CVA methods, threshold segmentation is performed on the difference image of change intensity. For the OB-HMAD method, the main MAD variable difference image is segmented. On the MAD variable difference image, changed objects will show lower brightness values, so changed objects can also be extracted through automated threshold

**Table 5.5** Accuracy evaluation results of multi-feature OBCD based on OB-HMAD and OB-MFF

| Model | Miss rate (%) | False alarm rate (%) | Overall accuracy (%) | Kappa |
|---|---|---|---|---|
| OB-HMAD | 21.5 | 12.3 | 81.6 | 0.653 |
| OB-MFF | 33.8 | 7.6 | 75.4 | 0.498 |

segmentation. For the OB-MFF method, it is determined by comparing the membership of each object to the changed and unchanged categories, and the calculation process of membership also uses an automated threshold method (as in Eq. (5.27)). The identification of change types is based on unsupervised classification of related statistics. Since neither OB-HMAD nor OB-MFF methods have direct statistics to judge the type of change, the change direction in CVA is combined to identify the type of change. The change detection results are shown in Fig. 5.16. It can be seen that the change detection results of OB-HMAD and OB-MFF are significantly different from those of SAW-CVA, and there are omissions and misjudgments. The error confusion matrix shows (Table 5.5) that the overall accuracy and Kappa coefficient of OB-HMAD and OB-MFF are lower than those of SAW-CVA.

### 5.5.2 Comparison of Different Threshold Segmentation Methods

According to the second principle of the theoretical framework, a threshold needs to be selected to distinguish between changed and unchanged objects, usually obtained through an automated threshold segmentation algorithm. Three commonly used threshold selection methods are compared with the EM algorithm to compare their selected threshold results and their time running costs. The Bhattacharyya distance is chosen to evaluate the separability of the changed and unchanged classes based on different threshold segmentation methods. 1. Maximum inter-class variance method The maximum inter-class variance method (Otsu) (Otsu, Otsu's method (1975) is the most commonly used global threshold automatic selection method in threshold algorithms, derived based on the least squares method.

For the difference image of CVA $f(x, y)$, assume the threshold $T$ divides the image into changed $O_c$ and unchanged $O_u$ categories, the probability of $O_c$ occurring is $\omega_c$, the average change intensity is $\rho_c$; the probability of $O_u$ occurring is $\omega_u$, the average change intensity is $\rho_u$, the total average change intensity of the difference image is $\rho_T$, then the inter-class variance is defined as

$$\sigma^2 = \omega_c \times (\rho_c - \rho_T)^2 + \omega_u \times (\rho_u - \rho_T)^2 \tag{5.33}$$

Let $T$ take values within a specified range, when $\sigma^2$ is at its maximum, the current $T$ is the appropriate threshold.

1. Improved Maximum Inter-class Variance Method

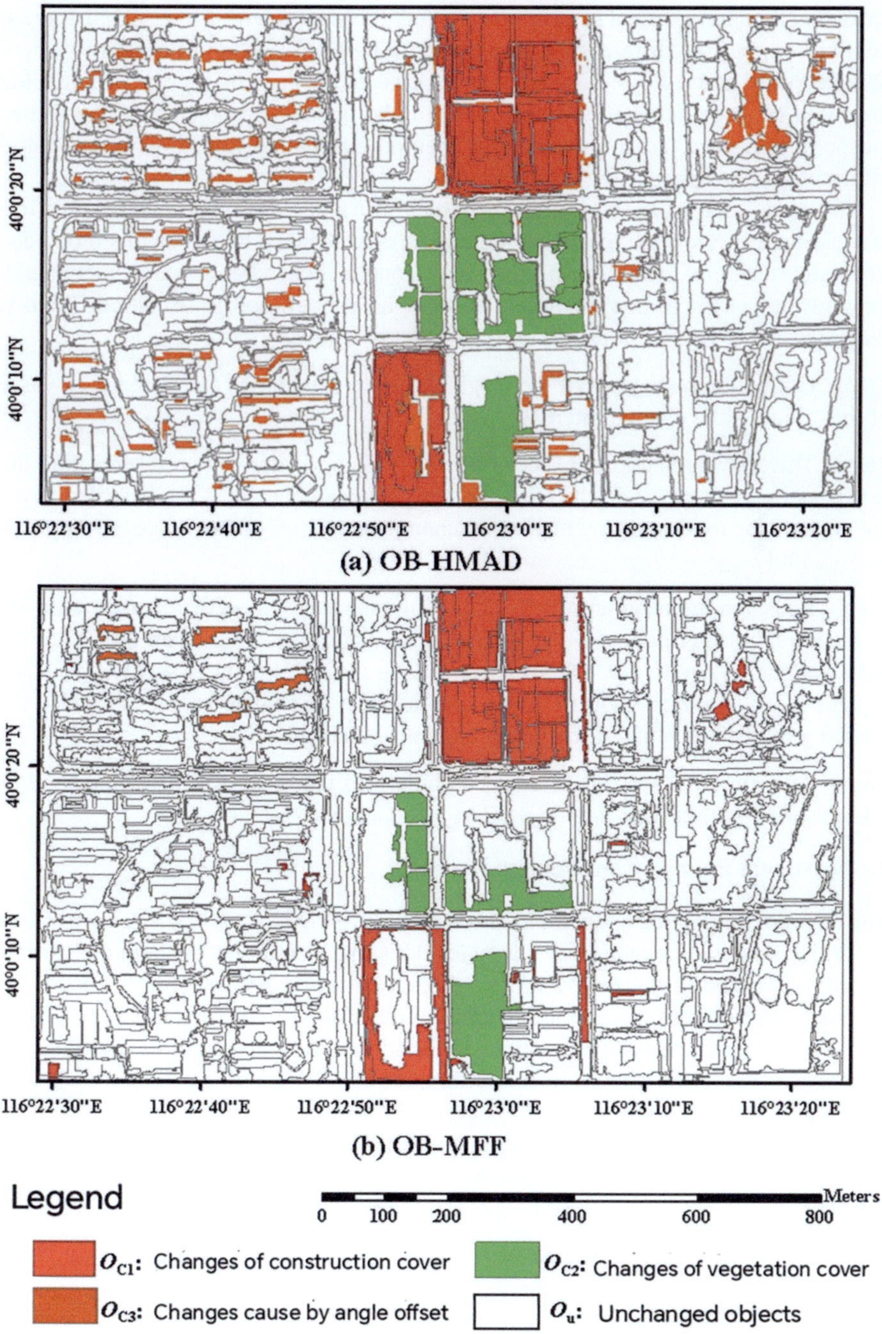


**Fig. 5.16** OB-HMAD and OB-MFF change detection results (colour map attached)

Otsu's method has an ideal segmentation effect for images with a bimodal histogram distribution, but the difference image obtained by CVA does not necessarily show obvious peaks and valleys. Simply using the change intensity histogram to obtain the threshold cannot guarantee that the changed and unchanged objects can be distinguished. Therefore, an improved Otsu method combining neighbourhood spatial information is proposed.

The basic idea of the algorithm is: firstly, a rough segmentation of the difference image is performed to extract the most likely changed and unchanged objects. Then, for the remaining objects that may contain both changes and unchanged, they are processed in conjunction with their neighbourhood information. Finally, the processed image is binarized using the Otsu method.

Step1: Obtain the histogram of the original difference image $f(x, y)$, the grey value of this graph is the change intensity value.

Step2: Traverse the histogram according to the change intensity grey level, find the first peak and the last peak, the corresponding change intensities are recorded as $\rho_0$ and $\rho_1$, and the minimum and maximum change intensities of the image are recorded as $\rho_{min}$ and $\rho_{max}$.

Step3: Reassign the original pixel's change intensity based on the change intensity of the neighbouring pixels. If $f(x, y) > \rho_{min}$ and $f(x, y) < \rho_{max}$, then use $f(x, y) = 0.25 \times (|f(x+1, y) - f(x, y)| + |f(x-1, y) - f(x, y)| + |f(x, y+1) - f(x, y)| + |f(x, y-1) - f(x, y)|)$ to update the intensity of change value, otherwise the intensity of change value of the pixel remains unchanged.

Step4: The image processed as above is binarised using the Otsu method, with $T$ taking values within $[2 \times \rho_0 - \rho_{min}, 2 \times \rho_1 - \rho_{max}]$.

2. Maximum Entropy Thresholding Method

The method of selecting a threshold based on the principle of maximum entropy uses the threshold to divide the image's grey level histogram into independent categories, thereby maximising the total entropy of each category. By selecting a threshold, each category obtains the maximum amount of information (Carvalho et al. 2011). In this book, it can be assumed that when the sum of the entropy of the changed and unchanged categories reaches its maximum, the threshold at this time is the optimal threshold.

If the grey level of the change intensity of the difference image is $L$, then the entropy values of the changed and unchanged categories corresponding to the threshold $T$ are $H_c(T)$ and $H_u(T)$:

$$H_c(T) = -\sum_{i=t+1}^{L-1} p_i \log_2 p_i \tag{5.34}$$

$$H_u(T) = -\sum_{i=0}^{T} p_i \log_2 p_i \tag{5.35}$$

**Table 5.6** Comparison of different thresholding methods

| Experiment group | Evaluation indicator | Otsu | Improved Otsu | Maximum entropy | EM |
|---|---|---|---|---|---|
| Experiment one | Change intensity threshold $T$ | 240 | 232 | 230 | 235 |
| | Overall accuracy (%) | 85.36 | 86.62 | 86.19 | 88.28 |
| | Running time (s) | 90.65 | 107.77 | 66.03 | 45.32 |
| Experiment two | Change intensity threshold $T$ | 618 | 615 | 595 | 600 |
| | Overall accuracy (%) | 83.96 | 84.63 | 85.08 | 86.03 |
| | Running time (s) | 84.66 | 95.46 | 53.94 | 32.01 |

$$p_i = \frac{N_i}{N} \tag{5.36}$$

where $p_i$ is the probability of the $i$th level of change intensity occurring; $N_i$ is the number of pixels at the $i$th level of change intensity; $N$ is the total number of pixels in the histogram. The optimal threshold $T$ is then

$$T = \arg\max\{H_c(T) + H_u(T)\} \tag{5.37}$$

3. Comparison of Different Thresholding Methods

The above three methods and the EM method are applied to the process of selecting the threshold of change intensity in SAW-CVA, and their threshold selection results, overall accuracy and running time are compared.

As can be seen from Table 5.6, the results of these thresholding methods are quite similar, with little difference in overall accuracy, and are generally applicable to this book. Since both Otsu and the improved Otsu are similar to exhaustive methods, their running times are much higher than other algorithms. The maximum entropy method often requires the support of prior knowledge, and the EM algorithm can be used for parameter estimation with missing samples, making it more universal.

## 5.6 Summary

This chapter mainly discusses multi-feature change detection based on SAW-CVA, including the construction of the theoretical framework for multi-feature change detection based on SAW-CVA, and the design of the "two-step" pattern for multi-feature change detection based on SAW-CVA. The experiment compares the similarities and differences of other traditional multi-feature change detection methods and analyses the differences of different threshold segmentation algorithms.

(1) Based on SAW-CVA, a theoretical framework for multi-feature change detection of image objects was constructed. To address the issue that standard CVA cannot highlight the dominant features of image objects and effectively distinguish different types of changes, the standard CVA was improved. The dominant features of each image object were determined according to the difference in the internal standard deviation of each image object, and an adaptive weight that changes with the dominant feature was assigned. Then, the weighted features were used as the input of CVA to calculate the change intensity and direction of each change vector, and express them in polar coordinates. The study shows that compared to the Cartesian coordinate system, the change vectors obtained based on SAW-CVA have higher separability in the polar coordinate system and are easier to determine their change intensity threshold. At the same time, the distribution of the change direction of each change vector also has clustering characteristics, which is easy to distinguish and identify.

(2) A "two-step" detection pattern for multi-feature change of high-resolution remote sensing image objects based on SAW-CVA was designed and developed. Firstly, the EM algorithm was used to obtain the change intensity segmentation threshold and extract the change objects. Then, the $K$ -means clustering algorithm was used to cluster the change direction and obtain the category division threshold of the change direction, achieving the distinction and identification of multiple types of changes. The experimental results verify that SAW-CVA has a lower omission rate and misjudgment rate, and a higher recognition accuracy. It has a higher ability to distinguish and identify multiple types of changes, and can detect not only changes in spectral properties but also changes in texture patterns.

# Chapter 6
# Conclusion and Outlook

This book discusses the theoretical framework and detection pattern of multi-feature change detection of high-resolution remote sensing image objects based on object-oriented, proposes an image segmentation optimisation method based on quality evaluation index and hierarchical parameter selection; designs a multi-feature selection process of image objects based on the genetic particle swarm optimisation algorithm combined with the mean variance ratio fitness function, obtains reliable multi-feature combinations of image objects; constructs a multi-feature change detection theoretical framework based on adaptive weight change analysis, and a "two-step" change detection pattern, achieves better change detection effects than conventional change detection methods, and realises the distinction and identification of multiple types of changes.

## 6.1 Research Conclusions

The change detection of high-resolution remote sensing image objects has always been a hot research field in geographical science. The application of multi-feature analysis and the identification of multiple types of changes are also scientific issues that scholars focus on, thus promoting the development of image segmentation, feature selection, and change detection techniques. Although many scholars advocate using the object-oriented image analysis approach to deal with high-resolution remote sensing change detection problems, so far, a complete theoretical framework has not been formed. Moreover, for the problems of multi-feature analysis and multiple types of change identification of image objects, no general detection pattern has been given. This book, based on the object-oriented research approach, with image objects as the basic processing and analysis unit, studies the image segmentation parameter selection, feature selection, and multi-feature change detection based



Q. Chen et al., *Multiple Feature Object-Based Change Detection Technology for Very High-Resolution Remote Sensing Image*, https://doi.org/10.1007/978-981-95-3917-8_6

on high-resolution remote sensing image objects, and achieves ideal experimental results. The main research conclusions of this book are as follows.

(1) Constructed an image segmentation parameter optimisation method based on the segmentation quality evaluation index and scale hierarchical screening, and evaluated the segmentation quality using the geometric/quantity consistency difference between the reference object and the segmented object.

The generation of image objects of reasonable size and boundary alignment is a prerequisite and guarantee for overall research. Multiscale segmentation is the main means of generating image objects, with the key being the optimal selection of segmentation parameters. Firstly, in the process of generating image objects based on multiscale segmentation algorithms, the size and boundary division of image objects have a significant impact on subsequent feature selection and change detection research. The reasonable selection of segmentation parameters is a necessary condition to meet the research requirements. Secondly, by maximising the internal homogeneity and external heterogeneity of image objects, a multi-band segmentation quality evaluation index, meanNSQI, is constructed. Combined with hierarchical screening, this can optimise the selection of multiscale segmentation parameters. Then, based on the analysis of the overlap relationship between the reference object and the corresponding segmented object, the geometric and numerical consistency differences between the reference object and the corresponding segmented object are evaluated. This can measure the quality of the segmented objects obtained based on the relevant segmentation parameters, and evaluate the optimal results relative to the segmentation parameters. Experimental results show that this method of optimising and evaluating segmentation parameters is applicable to high-resolution remote sensing images of various types and resolutions. The evaluation results of different land cover/use areas show significant differences, and the types of consistency differences also vary in different areas. In addition, the research also shows that within a certain segmentation scale range, the evaluation index of consistency differences maintains a stable correlation with the segmentation scale, is sensitive to changes in the segmentation scale, and further proves that the segmentation parameter optimisation evaluation based on consistency difference analysis is reliable. The overlap of multi-temporal images in the same area will not have a significant impact on the optimal results of segmentation parameters, and the segmented objects obtained in this way can maintain a relatively consistent boundary with the original single-frame image segmentation objects.

(2) The selection process of multiple features of image objects has been optimised using the genetic particle swarm algorithm, significantly improving the computational efficiency and reliability of the algorithm. The selected feature combinations fully reflect the diversity of multiple features.

The selection of multiple features of image objects can provide sufficient conditions to reflect the advantages of multi-feature analysis in object-oriented image analysis. Firstly, the Genetic Particle Swarm Optimisation algorithm (GPSO) is used as the search algorithm, combined with the fitness evaluation function of the Ratio of Mean to Variance (RMV), to design a multi-feature selection process based on GPSO. This is introduced into the multi-feature change detection (OBCD) research of high-resolution remote sensing image objects. The research shows that this method has a high degree of automation, can quickly and intelligently select feature combinations suitable for research purposes, and can greatly reduce subjective human interference in the feature selection process. Secondly, compared with other feature selection algorithms, the multi-feature selection algorithm based on GPSO has a stronger global search capability, can avoid local convergence problems, and has superior convergence capability. Finally, the sensitivity analysis results of the initial parameters of the GPSO feature optimisation selection algorithm show that the number of candidate features $M$ and the initial population size of the particle swarm $N$ will have a certain impact on the accuracy, convergence, and efficiency of the algorithm's OBCD. Compared with the JMD and NNC fitness functions, RMV has advantages in algorithm convergence and reliability, and is more suitable as the fitness function of the GPSO feature optimisation selection algorithm for OBCD.

(3) Based on the adaptive weight change vector analysis method, a theoretical framework for multi-feature change detection of image objects has been constructed, and a "two-step" detection mode has been designed to solve the key problem of multiple change type recognition.

A multi-feature change detection theoretical framework based on Self-Adaptive Weight Change Vector Analysis (SAW-CVA) has been constructed, which includes assigning adaptive weights to each feature, obtaining the change intensity and direction of the change vector, and its new expression form in the polar coordinate system. The research shows that compared with the Cartesian coordinate system, the change vector obtained based on SAW-CVA has higher separability in the polar coordinate system, and the direction of change is more concentrated, which is conducive to the extraction of change objects and the distinction of change types.

A "two-step" detection mode for multi-feature change detection based on SAW-CVA has been designed and developed: on the one hand, the EM algorithm can be used to obtain the segmentation threshold of change intensity; on the other hand, the $K$ -means algorithm can be used to perform cluster analysis on the direction of change. Experimental results also verify that under this "two-step" mode, SAW-CVA has higher change detection accuracy and has a higher ability to distinguish and identify multiple change types. It can not only detect changes in spectral properties but also detect changes in texture patterns. The theoretical framework and detection mode based on SAW-CVA have strong universality and compatibility. Similar multi-feature change detection algorithms (such as OB-HMAD and fuzzy fusion, etc.) can be expressed under this theoretical framework and can use the "two-step" mode to achieve multi-feature change detection. In addition, the application prospects of different threshold segmentation algorithms under this theoretical framework are basically similar, and their segmentation accuracy for change intensity is also basically similar, but the EM algorithm has the ability to analyse missing samples and higher efficiency and universality.

## 6.2 Innovation Points

This book closely revolves around the scientific issues related to multi-feature analysis and multi-change type recognition in high-resolution remote sensing image change detection. It provides an effective solution, guided by object-oriented image analysis, and comprehensively utilises key technologies such as multi-scale image segmentation, construction of segmentation quality evaluation indices, consistency difference analysis, genetic particle swarm multi-feature optimisation selection, adaptive weight change vector analysis, and expectation maximisation threshold segmentation. The focus is on exploring a theory and process of image analysis based on object orientation. The research results have significant theoretical and practical value for image segmentation, multi-feature selection, and change detection of high-resolution remote sensing images. The innovation of this book lies in systematically proposing methods and models suitable for solving key issues such as multi-feature analysis and multi-change type recognition in high-resolution remote sensing image object change detection.

(1) A multi-scale image segmentation parameter optimisation algorithm based on quality index combined with hierarchical screening is proposed, and a segmentation object quality evaluation method based on geometric and quantity consistency difference analysis is explored. This segmentation parameter optimisation and evaluation method can effectively avoid subjective interference in the parameter optimisation process and provide accurate and reasonable image objects for subsequent research.
(2) An image object multi-feature optimisation method based on genetic particle swarm optimisation algorithm is designed, and a convergence comparison analysis of this optimisation method is conducted. This highly automated and intelligent method can fully reflect and utilise the advantages of multi-features in object-oriented analysis, obtain highly separable features containing more change information, and provide reliable and efficient input for multi-feature change detection.
(3) Based on adaptive weight change vector analysis, a theoretical framework for image object multi-feature change detection is constructed, and a "two-step" detection model for multi-feature changes of high-resolution remote sensing image objects is developed. The expression method based on polar coordinates can intuitively discover the distribution of change vectors in different change type areas. Using adaptive weight assignment and threshold segmentation can achieve better change detection results than standard change vector analysis methods, and can distinguish and identify more change type change objects.

## 6.3 Research Prospects

There are still some issues that need to be further explored in this book, such as segmentation parameter optimisation, multi-feature optimisation evaluation, multi-feature change detection, and change detection accuracy evaluation methods.

(1) Evaluating segmentation quality based on the geometric and quantity consistency difference between the reference object and the corresponding segmentation object is very reasonable for a single image. However, for the superposition of multiple images, although it can be regarded as an image with more bands for segmentation, the use of reference objects for consistency difference evaluation in this case is a problem worth exploring. Because for two high-resolution images of the same area at different times, there must be differences in the external boundaries or internal features of some objects, the reference object is no longer the only accurate reference, and the impact of this difference on the evaluation results needs to be considered. In addition, as can be known from the research in Chap. 3, there are differences in the segmentation parameter optimisation evaluation results in different land cover/use areas. Since the research area selected in this book is small and the surface type is relatively simple, this difference can be ignored in the optimisation results of segmentation parameters; but for larger areas with more complex surface cover types, the impact of this difference needs to be considered.

(2) The evaluation of the genetic particle swarm feature optimisation result is based on its performance in actual applications, that is, the accuracy of its change detection is used to measure the reliability of its feature selection results. This can only indirectly evaluate the applicability and reliability of the method, ignoring the consideration of the correlation, separability, and feature contribution of the selected features. Therefore, it is possible to consider using spatial feature correlation analysis, independence analysis, and information degree measurement to evaluate the quality of the optimised feature subset. While ensuring the reliability of the feature subset, consider the relationship between its various features, further mine the advantages of multi-feature analysis, and avoid the impact of mutual interference between features. The specific implementation needs further in-depth research.

(3). Multi-feature change detection based on SAW-CVA can accurately extract change objects and distinguish change objects of different change types. According to the research in Sect. 5.4.4, in a relatively complex surface environment, this method will be interfered by noise such as similar shadows, geometric registration errors, shooting angle offsets, etc. The method adopted in this book is to regard it as a type of "pseudo change object", which can be distinguished by change direction classification. Although it can effectively avoid interference, it does not really eliminate the impact, because these pseudo change objects are often very fine objects and are easily misclassified into normal change objects. Therefore, for this problem, it is necessary to continue to find a more suitable method to eliminate this interference.

(4) The evaluation of change detection accuracy is achieved by constructing an error confusion matrix, using indicators such as misjudgement rate, omission rate, and overall accuracy as evaluation standards. This supervised evaluation method can directly assess the effectiveness of change detection methods, but essentially, it is still a pixel-based evaluation method, only considering the category information of the object and ignoring other features. According to the guidance of OBIA, all image processing, analysis, and evaluation methods should be based on image objects and fully consider the actual geographical significance and multi-features of image objects. Therefore, in addition to considering the category information of image objects, the shape, theme, edge, position and other features of image objects should also be considered, that is, a STEP matrix should be constructed to evaluate the effect of change detection. This method still needs further in-depth exploration.

# References


Agrawal RK, Bawane NG (2015) Multiobjective PSO based adaption of neural network topology for pixel classification in satellite imagery. Appl Soft Comput 28:217–225

Allen TR, Kupfer JA (2000) Application of spherical statistics to change vector analysis of Landsat data: southern Appalachian spruce-fir forests. Remote Sens Environ 74(3):482–493

Angeline PJ (1998) Using selection to improve particle swarm optimization. Anchorage, AK, pp 84–89

Anusha M, Sathiaseelan J (2015) Feature selection using K-means genetic algorithm for multi-objective optimization. Procedia Comput Sci 57:1074–1080

Ashoori S, Mohammadi S (2011) Compare failure prediction models based on feature selection technique: empirical case from Iran. Procedia Computer Science 3(1):568–573

Baatz M, Schäpe A (2000) Multiresolution segmentation: an optimization approach for high quality multi-scale image segmentation. Angewandte Geographische Informationsverarbeitung 7:12–23

Barnes CF, Fritz H, Yoo J (2007) Hurricane disaster assessments with image-driven data mining in high-resolution satellite imagery. IEEE Trans Geosci Remote Sens 45(6):1631–1640

Berrendero JR, Cuevas A, Torrecilla JL (2015) The mRMR variable selection method: a comparative study for functional data. J Stat Comput Simul 86(5):1–17

Blaschke T (2005a) Towards a framework for change detection based on image objects. Manufacturing Engineer 113:1–9

Blaschke T (2005b) Towards a framework for change detection based on image objects. Göttinger Geographische Abhandlungen 113:1–9

Blaschke T (2010) Object based image analysis for remote sensing. ISPRS J Photogramm Remote Sens 65(1):2–16

Blaschke T, Lang S, Lorup E et al (2000) Object-oriented image processing in an integrated GIS/ remote sensing environment and perspectives for environmental applications. Environ Inf Plan, Polit Public 2:555–570

Blaschke T, Hay GJ, Kelly M et al (2014) Geographic object-based image analysis—towards a new paradigm. ISPRS J Photogramm Remote Sens 87:180–191

Bourennane S, Fossati C, Cailly A (2010) Improvement of classification for hyperspectral images based on tensor modeling. IEEE Geosci Remote Sens Lett 7(4):801–805

Bovolo F, Bruzzone L (2007) A Theoretical framework for unsupervised change detection based on change vector analysis in the polar domain. IEEE Trans Geosci Remote Sens 45(1):218–236

Bovolo F, Bruzzone L, Marconcini M (2008) A novel approach to unsupervised change detection based on a semisupervised SVM and a similarity measure. IEEE Trans Geosci Remote Sens 46(7):2070–2082

Bovolo F, Marchesi S, Bruzzone L (2012) A framework for automatic and unsupervised detection of multiple changes in multitemporal images. IEEE Trans Geosci Remote Sens 50(6):2196–2212




Q. Chen et al., *Multiple Feature Object-Based Change Detection Technology for Very High-Resolution Remote Sensing Image*, https://doi.org/10.1007/978-981-95-3917-8

Brusco MJ (2014) A comparison of simulated annealing algorithms for variable selection in principal component analysis and discriminant analysis. Comput Stat Data Anal 77:38–53
Bruzzone L, Prieto DF (2000a) An adaptive parcel-based technique for unsupervised change detection. Int J Remote Sens 21(4):817–822
Bruzzone L, Prieto DF (2000b) Automatic analysis of the difference image for unsupervised change detection. IEEE Trans Geosci Remote Sens 38(3):1171–1182
Bruzzone L, Roli F, Serpico SB (2010) An extension of the Jeffreys-Matusita distance to multiclass cases for feature selection. IEEE Trans Geosci Remote Sens 33(6):1318–1321
Byrne T, Mccain A, Kretchun T (1980) Testing chrysanthemums for disease resistance. Calif Agric 34(11/12):14–15
Calderón R, Montes-Borrego M, Landa BB et al (2014) Detection of downy mildew of opium poppy using high-resolution multi-spectral and thermal imagery acquired with an unmanned aerial vehicle. Precision Agric 15(6):639–661
Canty MJ, Nielsen AA (2006a) Visualization and unsupervised classification of changes in multispectral satellite imagery. Int J Remote Sens 27(27):3961–3975
Canty MJ, Nielsen AA (2006b) Visualization and unsupervised classification of changes in multispectral satellite imagery. Int J Remote Sens 27(18):3961–3975
Carleer AP (2010) Assessment of very high spatial resolution satellite image segmentations. Photogramm Eng Remote Sens 71(11):1285–1294
Carvalho OA, Guimarães RF, Gillespie AR et al (2011) A new approach to change vector analysis using distance and similarity measures. Remote Sens 3(12):2473–2493
Carvalho Júnior OA, Guimarães RF, Gillespie AR et al (2011) A new approach to change vector analysis using distance and similarity measures. Remote Sens 3(12):2473–2493
Castilla G, Hay GJ (2007) Image objects and geographic objects. Springer Berlin Heidelberg, Berlin, Germany
Celik T, Ma KK (2010) Unsupervised change detection for satellite images using dual-tree complex wavelet transform. IEEE Trans Geosc Remote Sens 48(3):1199–1210
Chang-Yao W, Zheng-Jun L, Chun-Yan Y (2006) An experimental study on imaging spectrometer data feature selection and wheat type identification. Journal of Remote Sensing 10(2):249–255
Chapran J (2012) Biometric writer identification: feature analysis and classification. Int J Pattern Recognit Artif Intell 20(20):483–503
Chen XW (2003) An improved branch and bound algorithm for feature selection. Pattern Recogn Lett 24(12):1925–1933
Chen Q (2012) Remote sensing estimation of methane emissions from rice fields based on regression analysis and DNDC model. Beijing Normal University, Beijing
Chen KM, Chen SY (2002) Colour texture segmentation using feature distributions. Pattern Recogn Lett 23(7):755–771
Chen J, He C (2001) Land use/cover change dynamic monitoring based on change vector analysis (I)—determination method of change threshold. J Remote Sens 5(4):259–266
Chen B, Hong J (1997) The problem of optimal feature subset selection. J Comput Sci 20(2):133–138
Chen S, Zhao Y (1990) Remote sensing geoscience analysis. Surveying and Mapping Press, Beijing
Chen J, Chun Yang HE, Shi PJ et al (2001) Land use/cover change detection with change vector analysis (CVA): change magnitude threshold determination. J Remote Sens 16:307–318
Chen J, Gong P, He C et al (2003) Land-use/land-cover change detection using improved change-vector analysis. Photogramm Eng Remote Sens 69(4):369–379
Chen YH, Feng T, Shi PJ et al (2006) Classification of remote sensing image based on object oriented and class rules. Geomat Inf Sci Wuhan Univ 31(4):316–320
Chen Y, Miao D, Wang R (2010) A rough set approach to feature selection based on ant colony optimisation. Pattern Recogn Lett 31(3):226–233
Chen Q, Chen Y, Jiang W (2015) High-resolution remote sensing image change detection based on OB-HMAD algorithm and spectral features. Spectrosc Spectr Anal 35(6):1709–1714
Chen W (2007) Research on remote sensing application technology for earthquake damage assessment. Lanzhou Institute of Earthquake, China Earthquake Administration, Lanzhou

Chuang LY, Tsai SW, Yang CH (2011) Improved binary particle swarm optimisation using catfish effect for feature selection. Expert Syst Appl 38(10):12699–12707

Chuang LY, Chang HW, Tu CJ et al (2008) Improved binary PSO for feature selection using gene expression data. Comput Biol Chem 32(1):29–38

Civco DL, Hurd JD, Wilson EH et al (2002) A comparison of land use and land cover change detection methods. Washington DC, USA, p 21

Civicioglu P (2013) Backtracking search optimisation algorithm for numerical optimisation problems. Appl Math Comput 219(15):8121–8144

Civicioglu P, Besdok E (2011) A Conceptual comparison of the cuckoo-search, particle swarm optimisation, differential evolution and artificial bee colony algorithms. Artif Intell Rev 39(4):315–346

Clinton N, Holt A, Scarborough J et al (2010) Accuracy assessment measures for object-based image segmentation goodness. Photogramm Eng Remote Sens 76(3):289–299

Coppin PR, Bauer ME (1996) Digital change detection in forest ecosystems with remote sensing imagery. Remote Sens Rev 13(3–4):207–234

Coppin P, Lambin E, Jonckheere I et al (2004) Digital change detection methods in natural ecosystem monitoring: a review. Int J Remote Sens 25(9):1565–1596

Costa GAOP, Feitosa RQ, Cazes TB et al (2008) Genetic adaptation of segmentation parameters. In: Lecture notes in geoinformation & cartography, pp 679–695

Coulter LL, Hope AS, Stow DA et al (2011) Time-space radiometric normalisation of TM/ETM+ images for land cover change detection. Int J Remote Sens 32(22):7539–7556

Dams J, Dujardin J, Reggers R et al (2013) Mapping impervious surface change from remote sensing for hydrological modelling. J Hydrol 485(485):84–95

Dash M, Liu H (1997) Feature selection for classification. Intell Data Anal 1(s 1–4):131–156

Dempster AP, Laird NM, Rubin DB (1977) Maximum likelihood from incomplete data via the EM algorithm. J Roy Stat Soc 39(1):1–38

Deng JS, Wang K, Deng YH et al (2008) PCA-based land-use change detection and analysis using multitemporal and multisensor satellite data. Int J Remote Sens 29(16):4823–4838

Ding C, Peng H (2011) Minimum redundancy feature selection for microarray gene expression data. J Bioinform Comput Biol 3(3):185–205

Dong W, Shen Z, Cheng X (2016) Rapid earthquake disaster assessment method based on high-resolution remote sensing target feature library. J Earth Inf Sci 18(05):699–707

Doxani G, Karantzalos K, Strati MT (2012) Monitoring urban changes based on scale-space filtering and object-oriented classification. Int J Appl Earth Obs Geoinf 15:38–48

Doyle W (1962) Operations useful for similarity-invariant pattern recognition. J ACM 9(2):259–267

Drăguţ L, Tiede D, Levick SR (2010) ESP: a tool to estimate scale parameter for multiresolution image segmentation of remotely sensed data. Int J Geogr Inf Sci 24(6):859–871

Drăguţ L, Csillik O, Eisank C et al (2014) Automated Parameterisation for multi-scale image segmentation on multiple layers. ISPRS J Photogramm Remote Sens 88(100):119–127

Dronova I, Gong P, Wang L (2011) Object-based analysis and change detection of major wetland cover types and their classification uncertainty during the low water period at Poyang Lake, China. Remote Sens Environ 115(12):3220–3236

Du P, Liu S (2012) Remote sensing image change detection integrating multiple features. J Remote Sens 16(4):663–677

Eberhart RC, Shi Y (2001) Particle swarm optimisation: developments, applications and resources. Seoul, Korea, pp 81–86

Eismann MT (2012) Hyperspectral remote sensing. SPIE Press Bellingham, Washington DC, USA

Espindola GM, Camara G, Reis IA et al (2006) Parameter selection for region-growing image segmentation algorithms using spatial autocorrelation. Int J Remote Sens 27(14):3035–3040

Evans C, Jones R, Svalbe I et al (2002) Segmenting multispectral Landsat TM images into field units. IEEE Trans Geosci Remote Sens 40(5):1054–1064

Feng Z, Wu X (2011) Feature selection algorithm based on trace ratio criterion and +L-R method. Comput Eng 37(17):136–139

Gangkofner U, Weichselbaum J, Kuntz S et al (2010) Update of the European high-resolution layer of built-up areas and soil sealing 2006 with image 2009 data. USA, pp 44–49

Gao F, de Colstoun EB, Ma R et al (2012) Mapping impervious surface expansion using medium-resolution satellite image time series: a case study in the Yangtze River Delta, China. Int J Remote Sens 33(24):7609–7628

Gao YF, Li BQ, Cai YD et al (2013) Prediction of active sites of enzymes by maximum relevance minimum redundancy (mRMR) feature selection. Mol BioSyst 9(1):61–69

Ghamisi P, Couceiro MS, Benediktsson JA (2015) A novel feature selection approach based on FODPSO and SVM. IEEE Trans Geosci Remote Sens 53(5):2935–2947

Gibbons BJD, Chakraborti S (2010) Nonparametric statistical inference (3rd edn), volume 131 of Statistics: text books and monographs. Arch Biochem & Biophys 283(1):206–209

Gong P, Li X, Xu B (2006) Some research issues in the theory and application methods of high-resolution image interpretation. J Remote Sens 10(01):1–5

Hame T, Heiler I, San Miguel J (1998) An unsupervised change detection and recognition system for forestry. Int J Remote Sens 19(6):1079–1099

Hao M (2015) Research on remote sensing image change detection method based on spatial information accuracy enhancement. China University of Mining and Technology, Xuzhou

Haralick RM, Shanmugam K, Dinstein IH (1973) Textural features for image classification. IEEE Trans Syst Man Cybern B Cybern 6:610–621

Hay GJ, Blaschke T (2010) Special issue: geographic object-based image analysis (GEOBIA). Photogramm Eng Remote Sens 76(2):121–122

Hay G, Castilla G (2008) Geographic object-based image analysis (GEOBIA): a new name for a new discipline. Springer, Berlin, Germany

Hay GJ, Castilla G, Wulder MA et al (2005) An automated object-based approach for the multiscale image segmentation of forest scenes. Int J Appl Earth Obs Geoinf 7(4):339–359

Hay GJ, Blaschke T, Marceau DJ et al (2003) A comparison of three image-object methods for the multiscale analysis of landscape structure. ISPRS J Photogramm Remote Sens 57(s 5–6):327–345

He C, Wei A, Shi P et al (2011) Detecting land-use/land-cover change in rural–urban fringe areas using extended change-vector analysis. Int J Appl Earth Obs Geoinf 13(4):572–585

He C, Li J, Zhang J et al (2005) Dynamic monitor on urban expansion based on an object-oriented approach. Seoul, Korea, p 2850

Honeycutt CE, Plotnick R (2008) Image analysis techniques and grey-level co-occurrence matrices (GLCM) for calculating bioturbation indices and characterising biogenic sedimentary structures. Comput Geosci 34(11):1461–1472

Hu X, Tao CV, Prenzel B (2005) Automatic Segmentation of high-resolution satellite imagery by integrating texture, intensity, and colour features. Photogramm Eng Remote Sens 71(12):1399–1406

Hussain M, Chen D, Cheng A et al (2013) Change detection from remotely sensed images: From pixel-based to object-based approaches. ISPRS J Photogramm Remote Sens 80:91–106

Inbarani HH, Azar AT, Jothi G (2014) Supervised hybrid feature selection based on PSO and rough sets for medical diagnosis. Comput Methods Programs Biomed 113(1):175–185

Jain AK, Duin RPW, Mao J (2000) Statistical pattern recognition: a review. IEEE Trans Pattern Anal Mach Intell 22(1):4–37

Jia W, Wu Q, Zhang J et al (2012) Research on technical methods for managing rural collective land for construction based on high-resolution remote sensing images. Land Resour Informatiz 2(2):49–54

Jiang W, Chen Q, Guo J et al (2010) Wetland classification of multispectral remote sensing images based on HPSO and FCM. Spectrosc Spectr Anal 30(12):3329–3333

Jin YQ, Chen F, Luo L (2004) Automatic detection of change direction of multi-temporal Ers-2 Sar images using two-threshold Em and Mrf algorithms. Imaging Sci J 52(4):234–241

Jin S, Yang L, Danielson P et al (2013) A comprehensive change detection method for updating the National Land Cover database to circa 2011. Remote Sens Environ 132(10):159–175

Johnson RD, Kasischke ES (1998) Change vector analysis: a technique for the multispectral monitoring of land cover and condition. Int J Remote Sens 19(3):411–426

Johnson B, Xie Z (2011) Unsupervised image segmentation evaluation and refinement using a multi-scale approach. ISPRS J Photogramm Remote Sens 66(4):473–483

Kapur J, Sahoo PK, Wong A (1985) A new method for grey-level picture thresholding using the entropy of the histogram. Comput vis, Graph, Image Process 29(3):273–285

Kennedy J, Eberhart R (1995) Particle swarm optimisation. Perth, Australia, pp 1942–1948

Kira K, Rendell LA (1992) The feature selection problem: traditional methods and a new algorithm. San Jose, Canada, pp 129–134

Kohli D, Warwadekar P, Kerle N et al (2013) Transferability of object-oriented image analysis methods for slum identification. Remote Sens 5(9):4209–4228

Kononenko I (1994) Estimating attributes: analysis and extensions of RELIEF. Springer Berlin Heidelberg, Berlin, Germany

Kruskal WH, Wallis WA (1952) Use of ranks in one-criterion variance analysis. J Am Stat Assoc 47(260):583–621

Landgrebe D (1999) Information extraction principles and methods for multispectral and hyper-spectral image data. Inf Process Remote Sens 82:3–38

Lang S, Albrecht F, Kienberger S et al (2010) Object validity for operational tasks in a policy context. J Spat Sci 55(1):9–22

Leichtle T, Gei C, Wurm M et al (2017) Unsupervised change detection in VHR remote sensing imagery—an object-based clustering approach in a dynamic urban environment. Int J Appl Earth Obs Geoinf 54:15–27

Li X (2010) Research on weed recognition method based on feature optimisation and multi-feature fusion. Jiangsu University, Zhenjiang

Li H (2014) Research on high-resolution remote sensing image object classification method and its application in urban and rural planning monitoring. China Agricultural University, Beijing

Li Y, Yang H, Chen X (2008) Application of histogram fitting method based on EM and BIC to remote sensing change detection threshold determination. J Remote Sens 12(1):85–91

Li L, Shu N, Wang K et al (2014) Remote sensing image change detection method integrating multiple features. J Surv Mapp 43(9):945–953

Li J, Li X (2013) Digital image processing, 2nd edn. Tsinghua University Press, Beijing

Liang LI, Ning S, Kai W et al (2014) Change detection method for remote sensing images based on multi-features fusion. Acta Geodaetica Et Cartographica Sinica 43:945–953

Listner C, Niemeyer I (2011) Object-based change detection. Photogramm Remote Sens Geoinformation 4(2011):233–245

Liu J (2008) Research on remote sensing dynamic monitoring of mineral resource development and potential for landscape ecological reconstruction in Cangshan County. Shandong Normal University, Jinan

Liu Y, Wang G, Chen H et al (2011) An improved particle swarm optimisation for feature selection. J Bionic Eng 08(2):3396–3400

Liu Y, Bian L, Meng Y et al (2012) Discrepancy measures for selecting optimal combination of parameter values in object-based image analysis. ISPRS J Photogramm Remote Sens 68(1):144–156

Lobo A, Chic O, Casterad A (1996) Classification of Mediterranean crops with multisensor data: per-pixel versus per-object statistics and image segmentation. Int J Remote Sens 17(12):2385–2400

Lu D, Mausel P, Brondizio E et al (2004) Change detection techniques. Int J Remote Sens 25(12):2365–2401

Lu D, Moran E, Hetrick S (2011) Detection of impervious surface change with multitemporal Landsat images in an urban–rural frontier. ISPRS J Photogramm Remote Sens off Publ Int Soc Photogramm Remote Sens 66(3):298–306

Luo J, Wang Q, Ma J et al (2002) EM improved algorithm for maximum likelihood classification method of remote sensing images. J Surv Mapp 31(3):234–239

Lv Z (2014) Research on surface cover change detection method considering image spatial features. Wuhan University, Wuhan

Ma J (2006) Research on specific target change detection technology of panchromatic band high spatial resolution remote sensing images. Graduate School of Chinese Academy of Sciences (Institute of Remote Sensing Applications), Beijing

Macqueen J (1967) Some methods for classification and analysis of multivariate observations. Berkeley, USA, pp 281–297

Maeda EE, Arcoverde GFB, Pellikka PKE et al (2011) Fire risk assessment in the Brazilian Amazon using MODIS imagery and change vector analysis. Appl Geogr 31(1):76–84

Malila WA (1980) Change vector analysis: an approach for detecting forest changes with Landsat. Purdue University, USA, p 385

Marcano-Cedeño A, Quintanilla-Dominguez J, Cortina-Januchs MG et al (2010) Feature selection using sequential forward selection and classification applying artificial metaplasticity neural network. Glendale, USA, pp 2845–2850

Marpu PR, Neubert M, Herold H et al (2010) Enhanced evaluation of image segmentation results. J Spat Sci 55(1):55–68

Martha TR, Kerle N, van Westen CJ et al (2011) Segment optimisation and data-driven thresholding for knowledge-based landslide detection by object-based image analysis. IEEE Trans Geosci Remote Sens 49(12):4928–4943

Mei A (2001) Introduction to remote sensing. Higher Education Press, Beijing

Mekhmoukh A, Mokrani K (2015) Improved fuzzy C-means based particle swarm optimisation initialisation and outlier rejection with level set methods for MR brain image segmentation. Comput Methods Programs Biomed 122(2):266–281

Mitra P, Murthy CA, Pal SK (2002) Unsupervised feature selection using feature similarity. IEEE Trans Pattern Anal Mach Intell 24(3):301–312

Möller M, Lymburner L, Volk M (2007) The comparison index: a tool for assessing the accuracy of image segmentation. Int J Appl Earth Obs Geoinf 9(3):311–321

Moon TK (1996) The expectation-maximisation algorithm. IEEE Signal Process Mag 13(6):47–60

Moradi P, Gholampour M (2016) A hybrid particle swarm optimisation for feature subset selection by integrating a novel local search strategy. World Health Organ 43:117–130

Morchid M, Dufour R, Bousquet P-M et al (2014) Feature selection using principal component analysis for massive retweet detection. Pattern Recogn Lett 49:33–39

Myint SW, Gober P, Brazel A et al (2011) Per-pixel vs. object-based classification of urban land cover extraction using high spatial resolution imagery. Remote Sens Environ 115(5):1145–1161

Nackaerts K, Vaesen K, Muys B et al (2005) Comparative performance of a modified change vector analysis in forest change detection. Int J Remote Sens 26(5):839–852

Narendra PM, Fukunaga K (1977) A branch and bound algorithm for feature subset selection. Electron Lett c-26(9):917–922

Nazir M, Majid-Mirza A, Ali-Khan S (2014) PSO-GA based optimised feature selection using facial and clothing information for gender classification. J Appl Res Technol 12(1):145–152

Neri F, Tirronen V (2010) Recent advances in differential evolution: a survey and experimental analysis. Artif Intell Rev 33(1):61–106

Neubert M, Herold H, Meinel G (2007) Assessing image segmentation quality—concepts, methods and application. In: Lecture notes in geoinformation and cartography, pp 769–784

Nielsen AA, Conradsen K, Simpson JJ (1998) Multivariate alteration detection (MAD) and MAF postprocessing in multispectral, bitemporal image data: new approaches to change detection studies. Remote Sens Environ 64(1):1–19

Nina Z, Wang L et al (2008) A modified t-test feature selection method and its application on the HapMap genotype data. Genomics Proteomics Bioinform 5(3–4):242–249

Otsu N (1975) A threshold selection method from grey-level histograms. Automatica 11(285–296):23–27

Pal NR, Pal SK (1989) Entropic thresholding. Signal Process 16(2):97–108

Peng H, Long F, Ding C (2005) Feature selection based on mutual information: criteria of max-dependency, max-relevance, and min-redundancy. IEEE Trans Pattern Anal Mach Intell 27(8):1226–1238

Peng HY, Jiang CF, Fang X et al (2014) Variable selection for Fisher linear discriminant analysis using the modified sequential backward selection algorithm for the microarray data. Appl Math Comput 238(7):132–140

Prieto MS, Allen AR (2003) A similarity metric for edge images. IEEE Trans Pattern Anal Mach Intell 25(10):1265–1273

Prieto LB, Fernandez D (2000) A minimum-cost thresholding technique for unsupervised change detection. Int J Remote Sens 21(21):3539–3544

Qu X (2010) A feature selection algorithm based on conditional entropy. J Taiyuan Univ Sci Technol 31(5):413–416

Rakotomamonjy A (2007) Analysis of SVM regression bounds for variable ranking. Neurocomputing 70(7):1489–1501

Reddi S, Rudin S, Keshavan H (1984) An optimal multiple threshold scheme for image segmentation. IEEE Trans Syst Man Cybern 4(4):661–665

Ridler T, Calvard S (1978) Picture thresholding using an iterative selection method. IEEE Trans Syst Man Cybern 8(8):630–632

Sahoo P, Wilkins C, Yeager J (1997) Threshold selection using Renyi's entropy. Pattern Recogn 30(1):71–84

Schneider A (2012) Monitoring land cover change in urban and peri-urban areas using dense time stacks of Landsat satellite data and a data mining approach. Remote Sens Environ 124:689–704

Shima K, Todoriki M, Suzuki A (2004) SVM-based feature selection of latent semantic features. Pattern Recogn Lett 25(9):1051–1057

Shortridge A (2007) Practical limits of Moran's autocorrelation index for raster class maps. Comput Environ Urban Syst 31(3):362–371

Singh A (1989) Review article digital change detection techniques using remotely-sensed data. Int J Remote Sens 10(6):989–1003

Sun L, Wang X (2004) A feature selection method based on genetic operation and intra-class and inter-class distance criterion theory. Comput Eng Appl 40(21):178–181

Sun Z, Bebis G, Miller R (2004) Object detection using feature subset selection. Pattern Recogn 37(11):2165–2176

Tan M, Pu J, Zheng B (2014) Optimisation of breast mass classification using sequential forward floating selection (SFFS) and a support vector machine (SVM) model. Int J Comput Assist Radiol Surg 43(10):1058–1064

Tescher AG (1988) Automatic thresholding of grey-level pictures using 2-D entropy. Tech Symp 829:29–35

Tewkesbury AP, Comber AJ, Tate NJ et al (2015) A critical synthesis of remotely sensed optical image change detection techniques. Remote Sens Environ 160:1–14

Thapa RB, Watanabe M, Motohka T et al (2015) Potential of high-resolution ALOS–PALSAR mosaic texture for aboveground forest carbon tracking in tropical region. Remote Sens Environ 160:122–133

Thonfeld F, Feilhauer H, Braun M et al (2016) Robust change vector analysis (RCVA) for multi-sensor very high resolution optical satellite data. Int J Appl Earth Obs Geoinformation 50(August 2016):131–140

Tian J, Chen DM (2007) Optimisation in multi-scale segmentation of high-resolution satellite images for artificial feature recognition. Int J Remote Sens 28(20):4625–4644

Tokarczyk P, Wegner JD, Walk S et al (2015) Features, colour spaces, and boosting: new insights on semantic classification of remote sensing images. IEEE Trans Geosci Remote Sens 53(1):280–295

Tong H, Maxwell T, Zhang Y et al (2012) A Supervised and fuzzy-based approach to determine optimal multi-resolution image segmentation parameters. Photogramm Eng Remote Sens 78(10):1029–1044

Trelea IC (2003) The particle swarm optimisation algorithm: convergence analysis and parameter selection. Inf Process Lett 85(6):317–325

Trias-Sanz R (2005) A metric for evaluating and comparing hierarchical and multi-scale image segmentations. Seoul, Korea, pp 5647–5650

Ververidis D, Kotropoulos C (2008) Fast and accurate sequential floating forward feature selection with the Bayes classifier applied to speech emotion recognition. Signal Process 88(12):2956–2970

Walter V (2004) Object-based evaluation of LIDAR and multispectral data for automatic change detection in GIS databases. Istanbul, Turkey, pp 12–23

Wang P (2007) Research on image segmentation method based on multi-scale theory. University of Science and Technology of China, Hefei

Wang X, Yang J, Teng X et al (2007) Feature selection based on rough sets and particle swarm optimisation. Pattern Recogn Lett 28(4):459–471

Wang A, Wang S, Lucieer A (2010a) Segmentation of multispectral high-resolution satellite imagery based on integrated feature distributions. Int J Remote Sens 31(6):1471–1483

Wang GT, Liang WY, Cheng JL (2010b) Change detection method of multiband remote sensing images based on fast expectation-maximisation algorithm and fuzzy fusion. J Infrared Millimetre Waves 29(5):383–388

Wang S, Li D, Wei Y et al (2011) A Feature selection method based on fisher's discriminant ratio for text sentiment classification. Expert Syst Appl 38(7):8696–8702

Wang J, Qin Q, Zhao J et al (2015) Knowledge-based detection and assessment of damaged roads using post-disaster high-resolution remote sensing image. Remote Sens 7(4):4948–4967

Warner T (2005) Hyperspherical direction cosine change vector analysis. Int J Remote Sens 26(6):1201–1215

Wei HL, Billings SA (2007) Feature subset selection and ranking for data dimensionality reduction. Pattern Anal Mach Intell IEEE Trans on 29(1):162–166

Wen L, Wu H, Jiang F et al (2016) Establishment of remote sensing monitoring classification system and interpretation symbols for reclamation patches based on high-resolution remote sensing images. Remote Sens Land Resour 28(01):172–177

Willhauck G, Schneider T, de Kok R et al (2000) Comparison of object oriented classification techniques and standard image analysis for the use of change detection between SPOT multispectral satellite images and aerial photos. Amsterdam, Netherlands, pp 35–42

Witharana C, Civco DL (2014) Optimising multi-resolution segmentation scale using empirical methods: exploring the sensitivity of the supervised discrepancy measure Euclidean distance 2 (ED2). ISPRS J Photogramm Remote Sens 87(1):108–121

Xian G, Homer C (2010) Updating the 2001 national land cover database impervious surface products to 2006 using Landsat imagery change detection methods. Remote Sens Environ 114(8):1676–1686

Xiaolu S, Bo C (2011) Change detection using change vector analysis from Landsat TM images in wuhan. Procedia Environ Sci 11:238–244

Xu T, Peng D, Wang W (2006) Improved histogram equalisation algorithm. Ordnance Autom 25(7):58–59

Xuan G, Chai P, Wu M (1996) Bhattacharyya distance feature selection. Pattern Recogn 2:195–199

Yang H, Zhang S, Deng K et al (2007) Research into a feature selection method for hyperspectral imagery using PSO and SVM. J China Univ Min Technol 17(4):473–478

Yang H, Du Q, Chen G (2012) Particle swarm optimisation-based hyperspectral dimensionality reduction for urban land cover classification. IEEE J Sel Top Appl Earth Obs Remote Sens 5(2):544–554

Yang XS, Deb S (2010) Cuckoo search via levy flights. Mathematics 210–214

Yao X, Wang X, Zhang Y et al (2012) Review of feature selection methods. Control Decis 27(2):161–166

Yao D, Yang J, Zhan X (2014) Feature selection algorithm based on random forest. J Jilin Univ: Eng Ed 44(1):137–141

Yu L, Liu H (2004) Efficient feature selection via analysis of relevance and redundancy. J Mach Learn Res 5(12):1205–1224

Yu H, Zhang S, Kong B et al (2010) Optimal scale selection for object-oriented remote sensing image classification research. J Chin Image Graph 15(2):352–360

Yu L, Liu H (2003) Feature selection for high-dimensional data: a fast correlation-based filter solution. Washington DC, USA, pp 856–863

Yunhao C, Xiaobing LI, Jin C et al (2002) The change of NDVI time series based on change vector analysis in China, 1983–1992. J Remote Sens 6(1):12–18

Zhang L, Wu C (2017) Current status and prospects of multi-temporal remote sensing image change detection. J Surv Mapp 46(10):1447–1459

Zhang J, Yang G (2005) Automatic detection method of land use and cover change with single temporal remote sensing data. J Remote Sens 9(3):294–299

Zhang H, Fritts JE, Goldman SA (2008) Image segmentation evaluation: a survey of unsupervised methods. Comput vis Image Underst 110(2):260–280

Zhang H, Sun K, Li W (2014) Object-oriented shadow detection and removal from urban high-resolution remote sensing images. IEEE Trans Geosci Remote Sens 52(11):6972–6982

Zhao J, Zhang Z (2009) Ant colony feature selection method based on fuzzy rough set information entropy. Comput Appl 29(01):109–111

Zhu Z, Ong YS, Dash M (2007) Wrapper-filter feature selection algorithm using a memetic framework. IEEE Trans Syst Man Cybern B Cybern 37(1):70–76

MIX
Papier aus verantwortungsvollen Quellen
Paper from responsible sources
FSC® C105338

FSC
www.fsc.org

If you have any concerns about our products,
you can contact us on
**ProductSafety@springernature.com**

In case Publisher is established outside the EU,
the EU authorized representative is:
**Springer Nature Customer Service Center GmbH**
**Europaplatz 3, 69115 Heidelberg, Germany**

Printed by Libri Plureos GmbH
in Hamburg, Germany